£35·00
m u

**Books are to be returned on or before
date below**

Molecular Electro-Optics

PART 2

ELECTRO-OPTICS SERIES

Series Editor: Dr. Herbert Elion
Managing Director, Electro-Optics
Arthur D. Little, Inc.
Cambridge, Massachusetts

Volume 1. Molecular Electro-Optics: Part 1—Theory and Methods; Part 2—Applications to Biopolymers, *edited by Chester T. O'Konski*

Volume 2. Fiber Optics in Communications Systems, *by Glenn R. Elion and Herbert A. Elion*

Other volumes in preparation

Molecular Electro-Optics

PART 2

Applications to Biopolymers

Edited by Chester T. O'Konski

Department of Chemistry
University of California
Berkeley, California

MARCEL DEKKER, INC.   New York and Basel

Library of Congress Cataloging in Publication Data

Main entry under title:

Molecular electro-optics.

    (Electro-optics series ; v. 1)
    CONTENTS: pt.1. Theory and methods.--pt.2. Applica-
tions to biopolymers.
    1. Electrooptics. 2. Polymers and polymerization--
Electric properties. 3. Polymers and polymerization--
Optical properties.
QC673.M64   537.2'4   75-43047
ISBN 0-8247-6402-1

MARCEL DEKKER, INC.
270 Madison Avenue, New York, New York 10016

Current printing (last digit):
10 9 8 7 6 5 4 3 2 1

PRINTED IN THE UNITED STATES OF AMERICA

PREFACE

The molecules of a fluid respond to an applied electric field by orientation or deformation, and this produces optical anisotropies on a macroscopic scale. The magnitudes and speeds of these responses are related to the molecular charge distribution, directional polarizabilities, and reorientational rates. Thus, electro-optic measurements provide a general and powerful approach for the determination of structural characteristics of molecules.

The Kerr electro-optic effect (electric double refraction) was discovered in 1875, but the characterization of macromolecules by electro-optic effects began only in the late 1940s with the introduction of pulsed electric fields and oscillographic photorecording. This approach enabled direct measurement of relaxation times which relate to molecular rotational diffusion. Starting in 1950, many experimental contributions by this method have appeared. With the aid of interpretations based upon classical and quantum theory, these experiments have led to a better understanding of the electric orientation phenomena. In this monograph we focus on biopolymers, but small molecules and colloidal dispersions are also discussed. Methods of measurement, the theories of various effects, interpretation of data in terms of macromolecular electric properties and size and shape, and a variety of applications are included.

The interdependence of experiment and theory often can be seen in reviewing the past two decades of progress in this field. An example is the use of high fields for orienting macromolecules to the saturation limit to facilitate analysis of the Kerr

constant. This development in our Berkeley laboratory was
accompanied by the extension of the electric birefringence theory
to high fields, and this is relevant to understanding the propagation of intense laser beams (see Chap. 12, "Nonlinear Electro-
Optics"). The relaxation and saturation techniques have led to
new information about macromolecules, as discussed in Chap. 3 and
in Part 2. Highly monochromatic laser sources have facilitated
development of the electrophoretic doppler shift technique (Chap.
9). High intensity lasers will probably produce further extensions of technique and new discoveries.

The literature in this field is scattered in journals of
physics, chemistry, and biology of many lands. The fact that no
reference book encompassing both theory and applications of
electro-optic relaxation methods is available first prompted me to
consider a book on electro-optics in the early 1960s. In 1964
Marcel Dekker suggested that I write a book on the electric properties of macromolecules. I decided not to undertake that broad
a project but in 1970, after the stimulation of a professorship in
Uppsala which provided the incentive to organize some of the knowledge in this field, I proposed instead the present monograph.
Because there have been many new developments in the last decade,
it seemed best to arrange a collaboration. The authors of each
chapter were especially selected on the basis of their original
contributions and their command of the material in the respective
subtopics.

As can be seen from the chapter titles, the subject has been
divided according to the special interests of individually invited
contributors. In Part 1, "Theory and Methods," I introduce the
subject with a chapter outlining some history. The authors of the
remaining chapters present the theory of various effects, the
apparatus and methods for their measurement, and review illustra-
tive applications. It was decided to exclude from this volume
crystal electro-optics in order to keep the monograph to a reasonable size. An exception occurs in Chap. 12, where some of the

nonlinear effects in crystals are discussed to present the theory
comprehensively. Part 2 consists of eight chapters which deal
primarily with the data obtained on various types of biological
systems, and with their interpretation. It is our hope that with
this foundation many workers not yet fully aware of the great
potential of electro-optic relaxation methods as an alternative or
supplement to ultracentrifugation, light scattering, transport
measurements, optical spectroscopy, and dielectric dispersion will
be persuaded to consider this relatively recent but powerful and
general approach to studies of the properties of molecules of all
kinds.

Electro-optic researches in my laboratory have been pursued
by several students and visiting scholars, most of whom are among
the authors. The first Ph.D. student, Arthur Haltner, deserves
special mention not only for his interesting choice of subjects,
but also for excellence of experimental execution. I am grateful
for the interest of Sydney Fleming, who saw at an early date the
potential of the electric birefringence relaxation technique and
fostered early studies in industry.

I am happy to acknowledge the interest of persons who
attended the small seminar and the small class at Berkeley in 1965
and 1973, respectively, and the kind interest and stimulating discussions of Prof. Stig Claesson at Uppsala University in the
spring of 1970. I thank my recent associates in electro-optics
research, John W. Jost, Michael C. Kwan, and Lloyd S. Shepard for
helpful suggestions on a number of chapters in the book.

My greatest appreciation goes to all the authors who have
joined to create this new treatise. Many of their names appear in
Chap. 1, in which the early (and sometimes also quite recent)
discoveries are outlined. Because of the way that their contributions have fitted together, the whole of the enterprise has become
greater than the sum of its outstanding parts.

Chester T. O'Konski

CONTRIBUTORS OF PART 2

Grüler, Hans Clemens,* Department of Experimental Physics III, University of Ulm, Ulm, West Germany

Jost, John W.,** Department of Chemistry, University of California, Berkeley, California

Keynes, R. D., Physiological Laboratory, University of Cambridge, Cambridge, England

Maestre, Marcos F., Space Science Laboratory, University of California, Berkeley, California

O'Konski, Chester T., Department of Chemistry, University of California, Berkeley, California

Scheffer, Terry J.,*** Institut für Angewandte Festkörperphysik der Fraunhoder-Gesellschaft, Freiburg, West Germany

Shirai, Michio, Department of Chemistry, College of General Education, University of Tokyo, Tokyo, Japan

Squire, Phil G., Department of Biochemistry, Colorado State University, Fort Collins, Colorado

Stellwagen, Nancy C., 1409 Cedar Street, Iowa City, Iowa

Yoshioka, Koshiro, Department of Chemistry, College of General Education, University of Tokyo, Komaba, Meguroku, Tokyo, Japan

*Present affiliation: Institut für Angewandte Festkörperphysik der Fraunhoder-Gesellschaft, Freiburg, West Germany.

**Present affiliation: Union Oil Company of California, Union Research Center, Brea, California.

***Present affiliation: Brown Boveri Research Center, CH-5401 Baden, Switzerland.

CONTENTS

CONTENTS OF PART 1

Theory and Methods

Molecular Electro-Optics

PART 2

Chapter 15

ELECTRO-OPTIC DATA ACQUISITION AND PROCESSING

John W. Jost* and Chester T. O'Konski

Department of Chemistry
University of California
Berkeley, California

I. INTRODUCTION

An important advantage of using electro-optic relaxation
methods to characterize biopolymer systems is the intrinsic speed
of a measurement. Techniques for measuring the properties of
macromolecules such as column chromatography, electrophoresis, and
ultracentrifugation, for example, may require hours to make the
measurements. With electro-optics, measurements are usually

---

*Present affiliation: Union Oil Company of California, Union
Research Center, Brea, California

faster than the time it takes to bring the solution in contact
with the electrodes. When recording the signals photographically,
as described in Chap. 3 (Part 1), the rate-limiting steps in data
reduction are developing of the film and measurements from the
film to obtain a signal vs. time table. The plotting of the data
for analysis by graphical methods or punching of cards for computer analysis to obtain the desired parameters characterizing the
macromolecules that automated instruments can now be designed and constructed for routine, accurate recording and analysis of data.

The length of time required from execution of an experiment
to calculation of the desired parameters, while merely a hindrance
for basic studies, has been a more serious impediment in the
development of electro-optics as a versatile analytical tool.
Experimental techniques have progressed to such an extent since
the first pulsed electric birefringence began on solutions of
macromolecules that analytical instruments can now be designed and
constructed.

We visualize future applications of electro-optic methods for
studies of biological cells, chromosomes, DNA, and enzymes in
biological laboratories and in medical clinics. We also can fore-
see industrial chemical applications, e.g., pigment characterization, catalyst analysis, and polymer synthesis studies or control
processes. Such extensions of the method would be greatly
accelerated by more rapid methods of recording and analyzing the
data. Here we briefly review methods already used and in the
literature and report in greater detail some of the new recording
procedures which involve high-speed electronic sampling and digital
conversion, with the instrument connected directly to a fast digital
computer. We also discuss the computational techniques which are
appropriate to analysis of relaxation data. Because of the wide-
spread use of single-pulse or transient methods, and our own involv-
ment in its further development, we focus on this approach.

II.   RAPID METHODS FOR ESTIMATION OF RELAXATION TIMES

A.   Oscilloscope Observations of Decay Times

Visual inspection of the oscilloscope trace in a single-pulse
experiment, using the apparatus described in Fig. 3 of Chap. 3
(Vol. 1), can lead very quickly to an estimate of the order of
magnitude of the relaxation time.   If there is distribution of
relaxation times, this is also qualitatively indicated.   A prac-
ticed person viewing an oscilloscope trace can estimate the value
of the relaxation time of a simple exponential decay process to
about ±20%, by observing the time, t_1, which is required for the
signal to decay to 1/e of its original value.   Carefully examining a photograph, t_1 can be estimated to ±2% under favorable
conditions.   The intercept with the baseline of the line tangent
to the initial slope of the decay curve, gives another time, t_0,
which also equals the relaxation time for a single exponential
decay process.   With more than one relaxation time, the decay
times obtained by the two methods mentioned above will not be
equal, as can readily be seen from an inspection of the equation
for a discrete sum of relaxation times, viz,

$$S(t) = \sum_{i=1}^{N} s_i^0 \exp\left(\frac{-t}{\tau_i}\right) \tag{1}$$

where s_i^0 is the contribution to the observed signal $S(t)$ from
the $\underline{i}$th relaxation process, at $t = 0$, with relaxation time τ_i
in a discrete series of N separate processes.   Extending the
idea introduced above one can define a decay time t_2, as equal
to one-half the time required for the signal to decay to $1/e^2$ of
its value at the instant of removal of the field, and decay times
t_i which are 1/i of the time required for decay to $1/e^i$.   A

comparison of $t_0, t_1, \ldots, t_i$ gives a rough indication of the breadth of distribution of relaxation times [1,2].

B.   Two-Parameter Log Normal Distribution

A rapid procedure of analysis has been developed [3,4] based upon the assumption of a continuous logarithmic type distribution of relaxation times. The electro-optic signal decay, for a continuous distribution, may be written

$$S(t) = \int_0^\infty s(\tau) \exp\left(\frac{-t}{\tau}\right) d\tau \tag{2}$$

where $s(\tau)$ is the distribution function in τ space for the electro-optic signal at the instant of removal of the field. The following function was chosen

$$s(\tau) = C\tau^{-1} \exp\left\{\frac{-(\ln \tau - \ln \bar{\tau})^2}{2\sigma_\tau^2}\right\} \tag{3}$$

where C is a constant, σ_τ is the width parameter of the distribution and $\bar{\tau}$ is the electro-optic mean relaxation time defined by

$$\bar{\tau} = \frac{\displaystyle\int_0^\infty \tau s(\tau) \, d\tau}{\displaystyle\int_0^\infty s(\tau) \, d\tau} \tag{4}$$

This function can be reduced to a logarithmic normal distribution and has the convenient property that $\bar{\tau}$ remains constant, while the width parameter σ_τ is varied.

Normalized relaxation curves, or plots of $S(t)/S(0)$ vs. t/τ, can be calculated with a computer for various values of σ_τ. Then comparison of two points on the experimental decay curves (e.g., t_1 and $2t_2$ as defined above) with the calculated curves

is sufficient to determine $\bar{\tau}$ and σ_τ. This method requires
relatively quantitative measurements so a permanently recorded
trace is normally used. The relaxation time distribution function
can be transformed to a particle size distribution function with
the aid of a model or assumptions relating particle size to amplitude of signal and to the relaxation time [3,5-7].

C. Graphical Analysis

A more exact, but far slower, procedure consists of plotting
the logarithm of the signal vs. the time. If this plot can be
extended to sufficiently long times with adequate signal-to-noise
ratio, then the slope of the tangent line at long times will give
the longest relaxation time in a discrete series. Extrapolating
back to zero time and subtracting calculated signal values for the
longest relaxation time from the corresponding experimental signals at the same times gives a remainder which is due to other
relaxation times. The log of this remainder is then plotted and
the cycle is repeated to give the various amplitudes, s_i^0, and
the τ_i values, until the remainder is insignificant compared to
noise or other uncertainties. This method has been used for the
analysis of blood-tissue exchanges [8], in radioactivity decay
studies [9] and also in electro-optic studies [1,10].

D. Instrumental Aids

Long-persistent oscilloscope screens and storage
oscilloscopes are very helpful in applying visual methods. The
storage oscilloscopes developed in the 1960s utilized special
cathode-ray tubes which retained the signal trace electronically
(for example, Tektronix Type 549). In the 1970s transient recorders have been developed which sample the signals and hold the
samples in a digital memory device, playing them out to a regular

oscilloscope or to a chart (pen) recorder upon command (for example, Biomation Models 610B, 805, or 8100, and Inter-Computer Electronics, Inc., Model No. PTR-9200). An example of a chart record for a DNA signal using a Biomation 8100 obtained in our laboratory is shown in Fig. 1.

III.   ANALOG ANALYSIS OF RELAXATION TIMES

It is straightforward to electronically produce a signal which is the superposition of a discrete sum of relaxation times. A schematic diagram of a network for doing this is shown in Fig. 2. The output of this network, stimulated by a short pulse terminating with the applied field pulse, can be compared with that of the electro-optic signal by means of a differential ampli-fier and an oscilloscope. By suitable adjustments of the RC

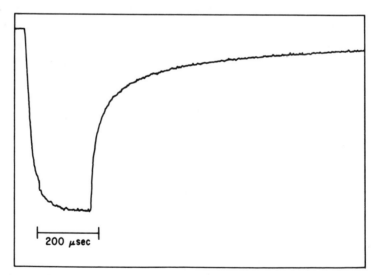

FIG. 1.   Birefringence signal from salmon sperm DNA (30 µg/cm^3 in 5×10^{-4} M Tris buffer) as recorded on a Biomation 8100 transient recorder. The orienting electric field was a 4 kV/cm dc pulse applied for 200 µsec. See Chaps. 3 and 18 for discussions of DNA birefringence.

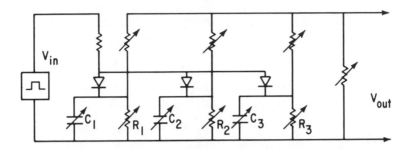

FIG. 2. Schematic of an analog synthesizer which will
generate $\sum_i s_i^0 \exp(-t/\tau_i)$; $\tau_i = R_i C_i$.

combinations and the values of R_0 to R_3, the synthesized
signal and electro-optic signal can be made to match over the entire decay region. Then the relaxation times can be obtained from
the RC time constants, and the relative amplitudes of the respective contributions from the relative values of R_1, R_2, and
R_3.

The disadvantage of this approach is that it requires many
trial and error adjustments and comparisons to achieve a match, as
well as requiring high accuracy of the overall system (better than
1%) to determine more than two relaxation times. Hence, although
a system like this was designed here in the 1950s, it was never
built. Recently, a similar system using operational amplifiers
was assembled and tests were reported [11]. This method is
relatively inexpensive and is of interest for insulating solutions
which give reproducible pulses. However, since it requires that
the signals be quantitatively reproducible over many pulses, which
is usually not the case with electrically conducting solutions, it
is not a promising approach for biopolymer solutions or other
aqueous colloids.

IV.   COMPUTATION OF ELECTRO-OPTIC RELAXATION TIMES

A.   Discrete Distributions

In this section we will discuss various numerical methods which have been used in the analysis of exponential decay curves. As was indicated in Sec. II, the graphical analysis of relaxation data can be a lengthy process. Also, if we wish to obtain reliable values for a distribution of relaxation times, an accurate method of analysis must be used. High-speed digital computers enable the rapid use of precise numerical methods to analyze electro-optic data. In Chap. 3 (Vol. 1) it was seen how the concentrations and sizes and shapes of rigid macromolecules can be obtained from a combination of saturation data and steady state birefringence values and from relaxation times. Flexible polymers were treated in Chap. 5 (Vol. 1). The reduced birefringence, $\Delta n(t)/\Delta n(0)$ [1,12] can be taken as the function $S(t)$ of Eq. (1). Then s_i^0 becomes the coefficient which measures the fractional contribution of the ith relaxation time to the total observed birefringence at $t = 0$, the instant of removal of the field, and τ_i is the ith relaxation time.

In the case of a continuous distribution given by Eq. (2) we may introduce $\alpha \equiv 1/\tau$. Then (2) becomes

$$S(t) = \int_0^\infty g(\alpha) \exp(-\alpha t) \, d\alpha \qquad (5)$$

which is of the form of a Laplace integral. The function $g(\alpha)$ gives the distribution of the relaxation times in α space. Methods for the analysis of experimental data in terms of a continuous distribution will be discussed in Sec. IVB.

1.   Least Squares Fitting of Data

Since the method of least squares plays such a central role in the applications to be discussed, we will first give a brief description of the general technique. A convenient criterion of

the goodness of fit of a set of experimental points to a
mathematical function having a number of adjustable parameters is
the quantity chi-squared defined by the equation

$$\chi^2 = \sum_{j=1}^{n} \frac{(x_j - \xi_j)^2}{\sigma_j^2} \tag{6}$$

where x_j is the value of the jth experimental point, ξ_j is
the corresponding computer point and σ_j is the standard error of
the jth experimental point. The smaller the value of χ^2, for
a given number of observations, n, the better the fit. Thus,
to obtain the best fit, we wish to minimize χ^2 with respect to
the adjustable parameters.

In least squares procedures, $\sum_j (x_j - \xi_j)^2$ often is minimized. This is equivalent to assuming that the σ_j^2 are constant
for all x_j. This would be true, for example, when the random
noise in the signal has an amplitude independent of its level. If
σ_j is a function of j, then an appropriate weighting function
can be introduced. One must analyze the nature of the experimental errors to arrive at this function. For computation of a single
relaxation time, the weighting function has recently been discussed for a number of computationally short least square methods
[13]. The analysis of data derived from counting experiments,
e.g., photon counting in fluorescence decay, requires that χ^2 be
minimized since σ_j^2 is dependent on the signal amplitude.

The minimization of χ^2 is done by requiring that
$\partial\chi^2/\partial A_m = 0$ for all A_m, where the A_m are the complete set of
adjustable parameters. Substituting for χ^2 and taking the
indicated derivatives, we find, for each A_m,

$$\frac{\partial \sum_{j=1}^{n} \frac{(x_j - \xi_j)^2}{\sigma_j^2}}{\partial A_m} = \sum_{n=1}^{n} \frac{-2(x_j - \xi_j)}{\sigma_j^2} \frac{\partial \xi_j}{\partial A_m} = 0 \tag{7}$$

The computed points ξ_j are given by the value of the fitting function at $t = t_j$ (taking t as the independent variable). The set of equations which result upon substitution of the theoretical fitting function into Eq. (7) will be exactly soluble only if the resulting set of equations is linear.

In the case of interest here the fitting function is a sum of exponentials and the resulting set of equations is not linear. However, a solution can be found by linearization of the problem. We consider first, then, the solution of the linear problem.

If the ξ_j can be written as linear functions of the A_m,

$$\xi_j = \sum_{m=1}^{N} C_{jm} A_m,$$ where C_{jm} depend only on the value of m and j and N is the number of adjustable parameters, then the minimization problem, Eq. (7), is reduced to

$$\sum_{j=1}^{n} \frac{(x_j - \sum_{\ell=1}^{N} C_{j\ell} A_\ell)}{\sigma_j^2} C_{jm} = 0 \tag{8}$$

or

$$\sum_{j=1}^{n} \frac{x_j C_{jm}}{\sigma_j^2} = \sum_{j=1}^{n} \sum_{\ell=1}^{N} \frac{C_{jm} C_{j\ell}}{\sigma_j^2} A_\ell \tag{9}$$

for all m. This set of N linear equations in N unknowns is referred to as "the normal equations." Their solution can be obtained formally by constructing the vectors

$$X = (X_m) = \sum_{j=1}^{n} \frac{C_{jm}}{\sigma_j^2} x_j \tag{10}$$

$$A = (A_m) \tag{11}$$

and the matrix

$$M = (M_{m\ell}) = \sum_{j=1}^{n} \frac{C_{jm} C_{j\ell}}{\sigma_j^2} \tag{12}$$

The set of simultaneous equations becomes the matrix

$$X = MA \tag{13}$$

with the solution

$$A = M^{-1}X \tag{14}$$

To solve the nonlinear curve fitting problem, the fitting function must be linearized and a solution to the resulting linear equations found. The usual method of linearization is by a Taylor series expansion of the fitting function.

In any case where the data are to be fitted to a function which is the sum of a product of terms consisting of an adjustable parameter and another function of another adjustable parameter (e.g., as a sum of exponentials), the calculated values may be written

$$\xi_j = \sum_{k=1}^{N/2} a_k f(b_k, t_j) \tag{15}$$

Here $f(b_k, t_j)$ indicates a function whose value depends both on the parameter b_k and the independent variable t_j. If the fitting function is a sum of exponential terms as in Eq. (1) above, then the a_k are the preexponential factors, the b_k are the relaxation times and $N/2$ is the number of relaxation times. The function $f(b_k, t_j)$ can be expanded as a Taylor series about a particular value of $b_k = b_k^0$. Rejecting higher terms, one obtains

$$f(b_k, t_j) = f(b_k^0, t_j) + (b_k^0 - b_k) \left. \frac{\partial f(b_k, t_j)}{\partial b_k} \right|_{b_k = b_k^0} \tag{16}$$

Substituting this for $f(b_k, t_j)$ in Eq. (15) gives

$$\xi_j = \sum_{k=1}^{N/2} a_k [f(b_k^0, t_j) + \Delta b_k f'(b_k^0, t_j)] \tag{17}$$

or $\quad \xi_j = \displaystyle\sum_{k=1}^{N/2} [a_k f(b_k^0, t_j) + a_k \Delta b_k f'(b_k^0, t_j)]$

It will be recognized that this is in the linear form given above, with

$$
\begin{aligned}
A_m &= a_k \text{ for } m = 1, \ldots, \frac{N}{2} \\
A_m &= a_k \Delta b_k \text{ for } m = \frac{N}{2} + 1, \ldots, N
\end{aligned}
\tag{18}
$$

$$
\begin{aligned}
C_{jm} &= f(b_k^0, t_j) \text{ for } m = 1, \ldots, \frac{N}{2} \\
C_{jm} &= f'(b_k^0, t_j) \text{ for } \frac{N}{2} + 1, \ldots, N
\end{aligned}
\tag{19}
$$

Thus, by the Taylor series expansion, we have reduced the nonlinear problem to a linear one. Inserting a set of initial b_k^0 into Eq. (17) and applying Eqs. (10) to (14), one may calculate for the a_k and Δb_k. The values of Δb_k are then used to provide new values $(b_k^0)' = b_k^0 + \Delta b_k$, and the calculation is iterated until the fit is satisfactory.

The iterative nonlinear least squares procedure has the hazard that, with a number of adjustable parameters, there may be a number of subsidiary minima, and one may find one of those. Another difficulty which is occasionally encountered is that certain combinations of coefficients lead to a singular M matrix, in which case no inverse exists. This means that for those values of the matrix elements the coefficients are not linearly independent. This is readily circumvented by appropriate programming.

All of the methods we will discuss, except one, require that the number of exponentials to be fit be fixed in advance. This raises the question of how best to decide this point. The most satisfactory solution is to repeat the calculation using varying numbers of exponentials and then deciding on the basis of the minimum value of χ^2. Since the Weierstrass approximation theorem [14] assures us that any well-behaved function can be approximated by a sufficiently long series of complete functions, we must also take into account Occam's "razor" in deciding the

number of exponentials required to fit the experimental data.
Occam's razor states that of two possible theories the simpler is
to be preferred. That is to say, the decrease in χ^2 must be
significant in order for a series with a greater number of terms
to be chosen as giving a better fit than one with fewer terms.
The significance of a decrease in χ^2 can be judged by using the
tables of the function $P(\chi^2 > \chi_0^2)$. (Commonly tabulated in, for
instance, the Handbook of Chemistry and Physics.) The function
$P(\chi^2 > \chi_0^2)$ gives the probability that for a given value of $n-N$,
the number of observations minus the number of parameters, which
is called the number of degrees of freedom, the value of χ^2
would be expected to be greater than that observed, χ_0^2. If the
value of $P(\chi^2 > \chi_0^2)$ is large then the fit is good. Thus if two
sets of exponentials give the same value for $P(\chi^2 > \chi_0^2)$ we are
not justified in saying one gives a better fit than the other,
even if the values of χ^2 differ. For a discussion of these
points see Mathews and Walker [15, p. 361].

2. Prony's Method

     This method is discussed by Lanczos [16], Buckingham [17],
and Perl [9]. The derivation and description that follows is
based on that in Practical Analysis by Willers [18].

     The problem we wish to solve is given by Eq. (1). If we
restrict ourselves to equally spaced data points, we can write
$t_m = t_1 + (m - 1)h$, where h is the separation between abscissa
values of the variable t, and we can then write the value of
$S(t)$ at any particular t_m as

$$
\begin{aligned}
S_m &= \sum_{i=1}^{N} s_i^0 \exp\left\{\frac{-t_1}{\tau_i} - \frac{(m-1)h}{\tau_i}\right\} \\
&= \sum_{i=1}^{N} s_i^0 \exp\left(\frac{-t_1}{\tau_i}\right) \exp\left\{\frac{-(m-1)h}{\tau_i}\right\} \\
&= \sum_{i=1}^{N} s_i^0 \exp\left(\frac{-t_1}{\tau_i}\right)\left\{\exp\left(\frac{-h}{\tau_i}\right)\right\}^{m-1}
\end{aligned}
\tag{20}
$$

Defining $u_i = \exp(-h/\tau_i)$ and $f_i = s_i^0 \exp(-t_1/\tau_i)$ the last equation becomes

$$S_m = \sum_{i=1}^{N} f_i u_i^{m-1} \tag{21}$$

For the case of three exponentials the first four data points give the system of simultaneous equations:

$$\begin{aligned}
S_1 &= f_1 + f_2 + f_3 \\
S_2 &= f_1 u_1 + f_2 u_2 + f_3 u_3 \\
S_3 &= f_1 u_1^2 + f_2 u_2^2 + f_3 u_3^2 \\
S_4 &= f_1 u_1^3 + f_2 u_2^3 + f_3 u_3^3
\end{aligned} \tag{22}$$

A system of simultaneous equations such as the above can be solved by converting to homogeneous form and setting the determinant of the coefficients equal to zero, or by multiplying the first three equations by $T_3 = -u_1 u_2 u_3$, $T_2 = u_1 u_2 + u_2 u_3 + u_3 u_1$, and $T_1 = u_1 + u_2 + u_3$, respectively, and then adding all four equations. (Note that the above problem is the solution of three simultaneous linear equations with a constraint.) The T_i's defined above are the symmetric functions of the unknowns u_1, u_2, and u_3 and are the coefficients in the cubic equation $x^3 + T_1 x^2 + T_2 x + T_3 = 0$, whose roots are the u_i. Performing the indicated multiplication and addition we obtain

$$S_1 T_3 + S_2 T_2 + S_3 T_1 + S_4 = 0 \tag{23}$$

This equation has three unknowns; we therefore need three such equations to determine the S_i's. Since the equation holds in general for any four equidistant points the other two equations required are readily obtained, and are

$$S_2 T_3 + S_3 T_2 + S_4 T_1 + S_5 = 0 \tag{24}$$

and

$$S_3 T_3 + S_4 T_2 + S_5 T_1 + S_6 = 0$$

The above set of equation can be solved by standard methods to determine the T_i's and the u_i's are found as the roots of the cubic equation. However, Eqs. (23) and (24) are true only if the S_m's are the ordinate values of the exact function. Perl [9] had very limited success using this method with increments of the observed quantity rather than the quantity itself. From his brief description of his application (p. 213) we find that the total number of points used was five, the minimum number of experimental points required to fit two exponentials. This, of course, restricted the efficiency of the method, since he did not use as much information as was available.

In the case of real data the right-hand side of Eqs. (23) and (24) is not equal to zero, but is instead equal to some small number ϵ_m. Thus, with more than the minimum number of data points, we must use the method of least squares. For a complete derivation, see the section in Willers [18].

Tests have been made in this laboratory by M. Kwan on a least squares computer program based upon Prony's method as outlined above. Precise input data synthesized for a series of three relaxation times were read in at 50 or 100 equally spaced points. The least squares procedure was applied with the criterion that the sum of the squares of the ϵ_m would be a minimum, which leads to an explicit solution for the three relaxation times. Then a linear least squares procedure was used to calculate the best values of the preexponential terms, consistent with the relaxation times found from the above. For reasons which are not entirely clear, very exact input data (5 to 7 significant figures) were required to give good fits to the relaxation times chosen to generate the synthetic input data, and less exact data gave inaccurate values. At the present time we are investigating the possibility that the difficulties mentioned are due to the computational method used in the computer program rather than being inherent in the

method. An extension of Prony's technique using the method of
moments has been presented by Isenberg and Dyson [19].

3. Nonlinear Least Squares Programs

There are many programs available which will fit a set of
experimental data to a nonlinear function. These programs use
various methods for the minimization. A common method is the
Taylor series expansion discussed in the previous section. For a
discussion of other methods see the book by Wilde [20]. A recent
article describing a general nonlinear least squares program has
been written by Powell and MacDonald [21]. Also useful as a
general introduction is an article by James [22]. In our laboratory we adapted a data analysis program, SFRENIC, obtained from
the Los Alamos laboratory of the University of California via
Professor L. Ruby. This program uses the Taylor series expansion
technique, and does a least square fit to a series of exponential
terms. It is incorporated in the larger program BIREL (for birefringence relaxation), which computes the birefringence vs. time
curve from the voltage vs. time curve digitized by a transient
recorder and stored on magnetic tape, then smoothes the data and
reduces the number of points to fit to the desired number of
exponential terms, calculates various statistical parameters, and
displays the calculated curve and the experimental curve in the
printed output. The latter can be converted to a CALCOMP plot if
desired. The initial guesses of the parameter values are either
supplied by the user or calculated from the experimental smoothed
curve by the program using certain simple rules.

Another program utilizing a Taylor expansion method is NLIN,
very similar to SFRENIC, but written at our University of
California, Berkeley Computer Center by R. M. Baer. It has been
used to compare numerical data analyses with the graphical
"peeling" method, and has been checked with SFRENIC. Data analyses of some results obtained by S. B. Hwang in this laboratory on a
biological membrane vesicle preparation are shown in Table 1. It

TABLE 1

(A) Comparison of Graphical Method and NLIN Computer Calculation

	Experiment 1		Experiment 2		Experiment 3	
	Graphical	NLIN	Graphical	NLIN	Graphical	NLIN
Number of exponentials	3	3	3	3	3	3
Root mean square deviation ($\times 10^3$)	1.75	0.422	0.562	0.404	0.663	0.394
s_1^0	0.0148	0.0134	0.0153	0.0140	0.0133	0.0077
τ_1	0.027	0.024	0.032	0.021	0.024	0.00045
s_2^0	0.0145	0.0151	0.0125	0.0139	0.0125	0.0171
τ_2	0.152	0.138	0.168	0.160	0.129	0.0879
s_3^0	0.0248	0.0256	0.0204	0.0203	0.0271	0.0282
τ_3	2.46	3.32	2.83	3.00	3.25	2.86

(B) Comparison of NLIN and SFRENIC

	Experiment 4		Experiment 5		Experiment 6	
	NLIN	SFRENIC	NLIN	SFRENIC	NLIN	SFRENIC
Number of exponentials	2	2	3	3	3	3
Number of iterations	4^a	6^b	14^a	7^b	6^a	10^b
Root mean square deviation ($\times 10^3$)	5.62	6.05	2.00	2.17	2.36	2.63
s_1^0	0.0113	0.0113	0.0076	0.0076	0.0050	0.0050
τ_1	0.1389	0.1389	0.0392	0.0392	0.0336	0.0336
s_1^0	0.0419	0.0419	0.0160	0.0160	0.0145	0.0145
τ_2	2.163	2.163	0.3125	0.3126	0.3282	0.3283
s_3^0	--	--	0.0327	0.0327	0.0335	0.0335
τ_3	--	--	2.122	2.122	3.098	3.098

[a] Iterations ceased when the fractional change in the sum of the squares of the deviations reached 10^{-5}.

[b] Iterations ceased when the fractional changes in all the parameters reached 10^{-6}.

can be seen that the graphical and computed results are quite
similar, and that the two independent sets of least-square computations agree very well for six different experiments.

Table 2 indicates results obtained using this program to
analyze data for four biologically interesting macromolecules:
salmon sperm DNA, DNA extracted from the head of T2 phages, T5
phage, a DNA-containing virus, and a preparation containing the
Ca^{2+} ATPase material from a lobster microsomal preparation supplied by David Deamer of the UC-Davis campus. The first three
columns of the table give the sample, the applied field, and the
steady state optical retardation. The transient recorders digi-
tize over 2000 points. The decay curves are usually represented
by 1000 to 1500 of these points. This number is large enough to
allow smoothing; this reduces the number of data points for fast
computations in the least squares portion of the program, and also
reduces noise. Usually around 100 points resulted after the s
smoothing process. The last four columns indicate the number of
exponential terms used to fit the data, the calculated values of
relaxation time, and the corresponding preexponential factor (in
percent of the total steady state signal) and the value of $\overline{\tau}_f$

The value of $\overline{\tau}_f$ is obtained from fitting to a sum of
exponential terms using the relation

$$\overline{\tau}_f = \frac{\sum_{i=1}^{N} s_i^0 \tau_i}{\sum_{i=1}^{N} s_i^0}$$

It represents the birefringence average relaxation time corresponding to the generated fitting function, after normalization of
the fit. (The fit may not give the full initial amplitude when
there are too few terms.) When 1-, 2-, and 3-term fits are compared, the change of $\overline{\tau}_f$ correlates with improvements of the fit
as seen in a comparison of experimental and calculated curves;
thus $\overline{\tau}_f$ is a convenient numerical substitute for the curves,
allowing for compact tabulations.

TABLE 2

Discrete Relaxation Analysis on Biopolymers Using BIREL

Sample	Pulse: E (kV/cm) (duration)	$-\delta^0$ (rad)	No. of terms	τ_i (μsec)	s_i^0 (%)	$\bar{\tau}_f$ (μsec)[b]
DNA of T2 virus (14 μg/ml in 1 mM Tris and 0.1 mM MgSO$_4$)	0.5 (1 msec)	2.2×10^{-3}	2	64 / 1404	38 / 69	928
	1.0 (1 msec)	3.1×10^{-3}	2	125 / 842	35 / 55	563
	1.2 (1 msec)	5.5×10^{-3}	2	183 / 705	46 / 50	455
	2.5 (1 msec)	13.2×10^{-3}	2	86 / 1126	60 / 38	489
DNA from salmon sperm (30 μg/ml in 5 × 10^{-4} M Tris buffer)	9.0 (0.3 msec)	0.105 FSD[a] = 6.1%	1	527	41	527
		FSD = 2.0%	2	36 / 1000	79 / 24	262
		FSD = 1.5%	3	32 / 172 / 1570	25 / 19 / 16	497
T5 virus (420 μg/ml in 10^{-3} M Tris and 10^{-4} M MgSO$_4$)	0.28 (0.8 msec)	0.011 FSD = 3.2%	1	185	90	185
		FSD = 3.1%	2	1920	2	214
Ca^{2+} ATPase (solubilized vesicles)	9.5 (0.05 msec)		2	37 / 654	42 / 51	

[a] FSD is the fit standard deviation, or the standard deviation between averaged input points and the exponential fitting function.

[b] Due to rounding of s_i^0 and τ_i decimals, $\bar{\tau}_f$ values are slightly different from those which may be computed from the above s_i^0 and τ_i values.

$\bar{\tau}_f$ will, in general, be different from $\bar{\tau}$, the birefringence average relaxation time computed from the experimental points, whenever the fit is not adequate, or when $\bar{\tau}$ is computed from truncated data. Thus the computed $\bar{\tau}$ value was too low for the data in Fig. 3 (226 μsec) indicating that the data should have been sampled down to baseline if $\bar{\tau}$ values are to be computed.

For the purposes of comparison here some of the data were calculated using 1, 2, and 3-term expressions. Figure 3(a), (b), and (c) are CALCOMP plots of the experimental and calculated points for salmon sperm DNA relaxation data following removal of a 9 kV/cm pulse, applied for 0.3 msec, which was sufficient to achieve steady state birefringence. They correspond to 1, 2, and 3-term fits, respectively. The various parameters are given in Table 2. Clearly the DNA of salmon sperm has a wide distribution of relaxation times; a 3-term fit is needed to represent the data well.

The DNA of T2 required at least two terms to give a fair fit. Further studies are in progress to examine the much broader distribution of relaxation times expected with such a high molecular weight DNA (ca. 10^8 daltons).

Figure 4 is a CALCOMP plot of the experimental and calculated points for the T5 phage data in Table 2 for a 1-term fitting function. The relaxation behavior of the T5 phage is well described by a single relaxation time, as can be seen either from the plot in Fig. 4 or the values in Table 2. For the Ca^{2+} ATPase data of Table 2, two terms gave a much better fit to the relaxation data than one.

We have recently investigated a technique due to Hooke and Jeaves [23] called "pattern search." This method does a trial and error search for that combination of parameters which minimizes the sum of the squares of the deviations. The advantages of this technique are that the computations required are simple and repetitive and are thus easily done on a small computer. We have for instance run a program using this technique on an XDS-910 with 8K of available core with no difficulty. On the other hand, it has required a large amount of modification to fit the program SFRENIC described above into the XDS-910. As a matter of fact, the pattern search technique leads to a program which could easily fit into a programmable desk calculator. Its main disadvantage is that it is relatively slow if the initial guesses are poor.

4. Fourier Transform Solution of the Laplace Integral Equation

Gardner et al. [24] have described a method for obtaining the relaxation frequency spectrum using Fourier transform techniques to solve the Laplace integral equation. The observed signal has been expressed in terms of a sum of exponentials in Eq. (1). There, in the limit as N becomes large, or if we regard the series as a Dirichlet series, $S(t)$ can be written as a Stieltjes integral,

$$S(t) = \int_0^\infty \exp(-\alpha t) \, dh(\alpha) \tag{25}$$

where $h(\alpha)$ is a step function. This can be rewritten in an equivalent form using a sum of δ functions, $g(\alpha)$, as

$$S(t) = \int_0^\infty \exp(-\alpha t) g(\alpha) \, d\alpha$$

A plot of $g(\alpha)$ vs. α will give a frequency spectrum if experimental values are used to define $S(t)$. A peak in this spectrum indicates the value of one of the α_i, while the height is proportional to s_i^0.

The integral equation which defines $S(t)$ can be solved by using a general method outlined by Titchmarsh [25] and in a treatment similar to that of Paley and Wiener [26].

If we transform variables from α and t to $\alpha = e^{-y}$ and $t = e^x$ we find,

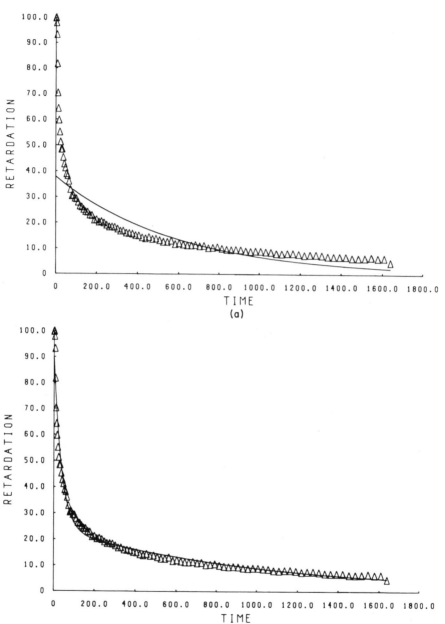

FIG. 3. Electric birefringence relaxation data for salmon sperm DNA as summarized in Table 2. (a) 1-term least squares fit. (b) 2-term fit. (c) 3-term fit. Optical retardations were normalized to 100 units, and time is in μsec. The triangles are smoothed data points and the curves are the fitting function. Obtained with the apparatus of Fig. 6. Clearly the 3-term fit is an improvement over 2 terms.

(Fig. 3 continued)

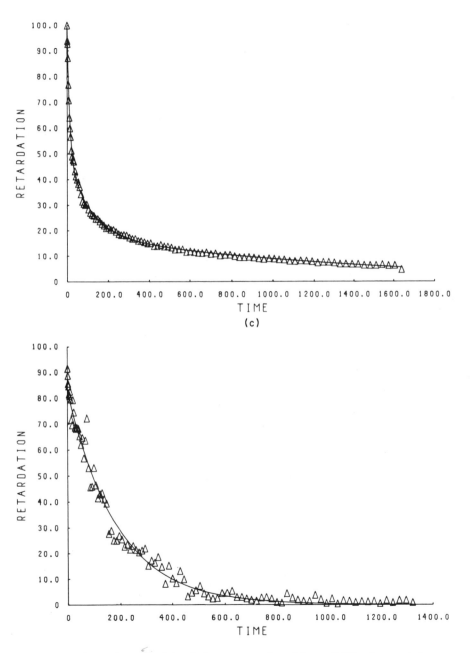

FIG. 4. Electric birefringence relaxation of T5 virus at low
fields, 0.1 kV/cm, with computations summarized in Table 2. Obtained
with the apparatus of Fig. 6. Time is in µsec.

$$S(e^x) = \int_{-\infty}^{\infty} \exp [-e^{(x-y)}] g(e^{-y}) e^{-y} \, dy \qquad (26)$$

Multiplying both sides by e^x we obtain

$$e^x S(e^x) = \int_{-\infty}^{\infty} \exp [-e^{(x-y)}] e^{(x-y)} g(e^{-y}) \, dy \qquad (27)$$

The Fourier transform of the left member of (27) is

$$F(\mu) = \sqrt{\frac{1}{2\pi}} \int_{-\infty}^{\infty} e^x S(e^x) e^{iux} \, dx \qquad (28)$$

and the Fourier transform of the right member is

$$F(\mu) = \sqrt{\frac{1}{2\pi}}$$

$$\int_{-\infty}^{\infty} \int_{-\infty}^{\infty} \exp [-e^{(x-y)}] e^{(x-y)} g(e^{-y}) \, dy \, e^{i\mu x} \, dx \qquad (29)$$

Changing variables, with $s = x - y$ we obtain

$$F(\mu) = \sqrt{\frac{1}{2\pi}} \int_{-\infty}^{\infty} \int_{-\infty}^{\infty} \exp [-e^s] e^s g(e^{-y}) \, dy \, e^{i\mu(s+y)} \, ds \qquad (30)$$

which can be rearranged to give

$$F(\mu) = \sqrt{\frac{1}{2\pi}} \int_{-\infty}^{\infty} g(e^{-y}) \exp (i\mu y) \, dy$$

$$\int_{-\infty}^{\infty} \exp (-e^s) e^s \exp (i\mu s) \, ds$$

At this point we note that $g(e^{-y}) \, dy = -(g(\alpha)/\alpha) \, d\alpha$. Thus if we can obtain $g(e^{-y})$ as a function of y we will have $g(\alpha)/\alpha$ as a function of α. The right-hand side of Eq. (31) is the product of two Fourier transforms, that of $g(e^{-y})$ and of $\exp (-e^s) e^s$. If we call these two Fourier transforms $G(\mu)$ and $K(\mu)$, respectively, we find that $F(\mu) = \sqrt{2\pi} G(\mu) K(\mu)$, which can be solved to give

$$G(\mu) = \overline{\frac{1}{2\pi}} \frac{F(\mu)}{K(\mu)} \tag{32}$$

If we take the inverse Fourier transform of $G(\mu)$ we obtain

$$g(e^{-y}) = \frac{1}{2\pi} \int_{-\infty}^{\infty} \frac{F(\mu)}{K(\mu)} e^{-iy\mu} d\mu \tag{33}$$

The integral $K(\mu)$ can be evaluated analytically and is the Euler integral for the complex Γ function,

$$K(\mu) = \overline{\frac{1}{2\pi}} \Gamma(1 + i\mu) \tag{34}$$

The determination of $g(e^{-y})$ as a function of y is thus reduced to finding the inverse Fourier transform of $F(\mu)$, which is the Fourier transform of the experimental data (with a change of variable), divided by the complex Γ function.

The chief attraction of this method is that there is no need to determine the number of exponentials to be fit. The number of exponentials is determined from the results of the calculation, in addition to the values of τ_i and s_i^0.

Gardner et al. have applied this technique to artificial data generated by combining various exponential terms and adding in "random" error to obtain realistic data. Results are obtained as values of the function $g(e^{-y})$ for various values of y. If these are plotted versus e^{-y} the resulting graph gives $g(\alpha)/\alpha$ as a function of α. Six such plots are given in their paper. In carrying out the integration of Eq. (33), it is necessary to set finite integration limits. The effects of varying this "cutoff" limit on the computed distribution are discussed in the original paper.

    B.   Continuous Distributions of Relaxation Times

1.   General

    A solution of macromolecules being investigated may be
polydisperse, may contain a very large number of components of
different molecular weights and different rotational diffusion
constants.   This situation can be represented by a continuous
distribution of relaxation times, and Eq. (5) replaces Eq. (1).
We can modify Eq. (5) slightly to obtain

$$S(t) = \int_{\alpha_0}^{\infty} \alpha g(\alpha) \exp(-\alpha t) \, d\alpha \tag{35}$$

The quantity α_0 is the reciprocal of the longest relaxation time
which contributes to the observed signal and can often be set
equal to zero.   The function $g(\alpha)$ gives the spectrum of rotational diffusion constants or relaxation frequencies and is
normalized as follows:

$$\int_{\alpha_0}^{\infty} g(\alpha) \, d\alpha = 1 \tag{36}$$

2.   Expansion Methods

    The integral expression for $S(t)$ given in Eq. (35) is the
Laplace transform of $h_{\alpha_0}(\alpha)g(\alpha)$, where $h_{\alpha_0}(\alpha)$ is the unit step
function with step at $\alpha = \alpha_0$.   If $g(\alpha)$ is expanded in a suitable series of functions the Laplace integral can be evaluated,
usually from available tables, and the data fit by adjustment of
the expansion parameters.   Matsumoto et al. (7) chose the expansion

$$g(\alpha) = \sum_{i=1}^{N} A_i (\alpha - \alpha_0)^i \exp(-B\alpha) \tag{37}$$

where A_i and B are parameters and the i's are integers. When this is substituted for $g(\alpha)$ in the Laplace integral we have

$$S(t) = \int_{\alpha_0}^{\infty} \sum_{i=1}^{N} A_i (\alpha - \alpha_0)^i \exp(-B\alpha) \exp(-\alpha t) \, d\alpha$$

$$= \int_{0}^{\infty} \sum_{i=1}^{N} h_{\alpha_0}(\alpha) A_i (\alpha - \alpha_0)^i \tag{38}$$

$$\exp(-B)(\alpha - \alpha_0 + \alpha_0) \exp(-\alpha t) \, d\alpha$$

The unit step function leads to a so-called "delaying" factor (see, for example, Kreider et al. [14, p. 195] in the transform. In general, if

$$f(t) = h_a(t) g(t - a) \tag{39}$$

$$L[f(t)] = \exp(-as) L[g(t)] \tag{40}$$

where L is the Laplace transform operator.

This allows us to write

$$S(t) = \exp(-\alpha_0 t) \sum_{i=1}^{N} \int_{0}^{\infty} A_i (\alpha)^i \tag{41}$$

$$\exp(-B\alpha - B\alpha_0) \exp(-\alpha t) \, d\alpha$$

The Laplace transform of the function

$$\frac{t^{r-1}}{(r-1)!} \exp(at) \; ; \; r > 0 \tag{42}$$

where r is an integer and a is a constant, is

$$\frac{1}{(s-a)^r} \quad \text{for } r > 0 \tag{43}$$

Applying this result we obtain

$$S(t) = \exp\ -(B + t)\alpha_0\ \sum_{i=1}^{N}\ \frac{A_i i!}{(B + t)^{i+1}} \tag{44}$$

The data can be fit to this function using the method of linear least squares outlined above if the values of α_0 and B are assigned initially. One can do a trial-and-error search for the values of α_0 and B which lead to the best values of χ^2, finding the best values of the A_i in each instance.

Since the distribution function $g(\alpha)$ is defined in the interval $(0,\infty)$ and is expected to be at least piecewise continuous, a natural set of functions to use for expansion are the Laguerre polynomials

$$L_n(x) = e^x\ \frac{d^n}{dx^n}\ \frac{x^n e^{-x}}{n!} \tag{45}$$

Methods for the inversion of the Laplace transform similar to this have been discussed by Papoulis [27]. His discussion is concerned with numerical inversion by orthogonal expansion methods. He discusses three methods, one of which involves expansion in terms of Laguerre polynomials. A more recent discussion and review of the numerical inversion of the Laplace transform for energy transport by a light pulse has been given by Renken and Biggs [28].

Matsumoto, Watanabe, and Yoshioka have applied the method described above to the calculation of $g(\alpha)$ for poly-L-glutamic acid (PLGA) in methanol [7] and methanol-water mixtures [6] to obtain a distribution in length of the macromolecules from the calculated $g(\alpha)$.

Figure 5(a) presents a calculated distribution curve obtained for a 1.0×10^{-3} solution of poly-L-glutamic acid (PLGA) in methanol. The birefringence decay curve is shown in Fig. 5(b). The parameters used in the expansion are given in the figure. This distribution can be interpreted in terms of the length

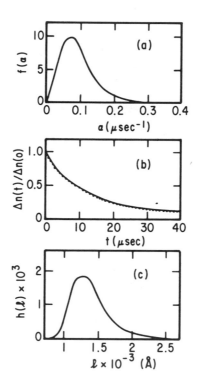

FIG 5. (a) Distribution function f(α) versus α, for PLGA in methanol, obtained from (b) by use of an approximating function. The parameters and coefficients are: $\alpha_0 = 0$, B = 45; A = 1.524 × 10², A = -6.299 × 10³, A = 7.255 × 10⁵. σ = 3.314 × 10⁻³. (b) Normalized birefringence, Δn(t)/Δn(0), of PLGA in methanol after removal of an electric field, as a function of time. Concentration, 1.0 × 10⁻³ g/cm³; temperature, 25°C; field strength, 5.0 × 10³ V/cm. (c) Length distribution, h(ℓ) vs. ℓ, for PLGA in methanol, obtained from (a).

distribution of the macromolecules. The relation obtained by
Matsumoto et al. is that $H(\ell)$, the distribution in length is
given by

$$h(\ell) = \frac{-\alpha g(\alpha)}{C\ell^4} \quad \frac{d \ln \alpha}{d \ln \alpha} \tag{46}$$

where C is found by requiring $\int_0^L h(\ell)d\ell = 1$. Figure 5(c) shows
the length distribution obtained using the data of Fig. 5(a), (b).

V.   COMPUTERIZED ELECTRO-OPTIC RELAXATION APPARATUS

The previous section has described a number of numerical techniques for extracting relaxation time distributions, buth discrete
and continuous, from the experimental measurement of a voltage transient which constitutes the electro-optic signal. The most time-
consuming step in going from the electro-optic signal to the relaxation times was the conversion of the photographed oscilloscope trace
to a voltage vs. time table. In order to eliminate this step and to
allow truly high speed data analysis the analog voltage signal can be
converted electronically to a series of digital signals.

Figure 6 is a schematic diagram of a computerized system which
was assembled and used here in 1973-74. A transient recorder of the
type mentioned in Sec. II was interfaced to an XDS-910 computer system assembled by W. D. Gwinn and collaborators [29]. The transient
recorders used were: Biomation 610B, 8100 and the Inter-Computer
Electronics PTR 9200. The recorder contains its own memory so that
data transfer to the computer can occur at whatever rate is convenient
up to some maximum rate which depends on the device in question. A
disadvantage of such a system is that the recorders presently available have a resolution of only 8 bits, that is, each voltage is converted to an 8-bit binary number. This resolution is equivalent to,
0.4% of full scale, which is in most cases better than other experimental uncertainties. However, data point coadding can be used to
improve the resolution if the solution under study is not degraded
by repeated applications of the Kerr pulse. Coaddition has been
used by Cohen et al. [30] to study the squid giant axon.

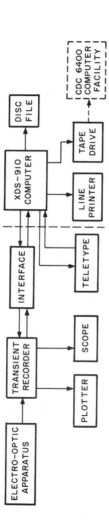

FIG. 6. Intermediate data acquisition system using a transient recorder, shared on-line computer, and separate data analysis facility.

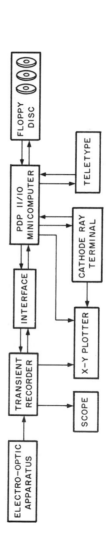

FIG. 7. Current data acquisition facility using a transient recorder and a dedicated minicomputer system.

Using the instrument shown in Fig. 6, data were stored by the XDS-910 on magnetic tape, then analyzed with an appropriate program (BIREL) on the UC Computer Center CDC 6400. This system was used to study various T-bacteriophage particles [31].

Our current system is shown in Fig. 7. It can handle fast relaxation times and also can be used to do signal averaging. The electro-optic apparatus is essentially that described in Chapter 3 of Part 1 of this treatise. We have utilized an Intercomputer Electronics PTR 9200 transient recorder for fast signals and dual channel recording; it has a 100 MHz maximum conversion rate. For slower signals (5 MHz maximum conversion rate) we used a Biomation 805. The ICE has 2816 8-bit words and the Biomation has 2024 8-bit words. Raw data points are transferred to the floppy discs, and can be plotted on the CRT with program PLOT which connects consecutive data points. To process data files the user examines on the CRT a display of numbers corresponding to signal amplitudes for consecutive samples and enters the "field-on" and "field-off" data point indices and manual instrument readings. Program SMOOTH is used to compute prepulse and postpulse baseline, amplitudes, standard deviations, and retardations from the raw data. Program AVETAU is useable directly on the raw data to calculate the average birefringence relaxation time. Program SMOOTH also preprocesses the data for non-linear least squares fitting, producing "smoothed" data files for both the pulse-on and the field-free decay periods. It also computes maximum retardation value and the value just before removal of the field.

The "smoothing" procedure consists of taking for each file the first 20 data points as they are recorded, then the next 60 are averaged three at a time to give 20 more "smoothed" points, then the next 140 are averaged seven at time, etc., up to 31 at a time (maximum averaging) and this reduces the data files to 100 to 120 points, typically, for more modest length for nonlinear least square fitting procedures. This strategy (increasing the number of points averaged as 2^n-1 as the signal amplitude falls exponentially) records the rapidly breaking start of a transient at highest resolution, then averages data points in the tail of the curve to improve S/N at lower amplitudes, and works well.

Typically, 1-3 exponential terms (2 to 6 independent parameters) are then fitted to the signal decay data. The signal buildup data may be fitted to a multiple term fitting function with both buildup and delay exponentials and a constant (typically 5 or 7 parameters for PM2 viral or calf thymus DNA, for example. Comparisons of the "smoothed" data points are then made with the computed fitting functions using either a linear (EOPLOT) or a logarithmic (LOGPLT) display on the CRT; the latter is far better. Another plotting program permits viewing the buildup data points and their fitting function. Inked data traces can be made on the X-Y plotter using a specially designed interface. With only 8K of computer core, a single iteration requires 12 seconds for the six independent parameter nonlinear fitting, and convergence (to within 1% for all parameters) occurs in 5-7 iterations, typically. Statistics and rounded data are printed out on the teletype when desired.

For a borader viewpoint, let us briefly consider the relationship between the relaxation methods we have been discussing and those encountered in the analysis of data from pulsed NMR and optical interferometry experiments. In these two experiments one also obtains exponential-type decay curves; however, species with different relaxation times tend to have different characteristic frequencies. Thus, when the data are Fournier transformed, the result is a series of Lorentzian lines centered at various frequencies. See, for example, the books by Farrar and Becker [31] and Bell [32] for a thorough introduction to these techniques. The ability to distinguish a line from an adjacent line is directly related to the separation in frequency between the two lines. In the electro-optic experiment the Fournier transform of the data gives a sum of Lorentzians, all centered at zero frequency. The decomposition of such a line shape is a much more difficult problem than the related problem encountered in NMR or optical interferometry. We note that if Prony's method (Sec. IVA2) is applied to an exponential decay curve with "wiggles", an imaginary root is found which corresponds to the sinusoidal behavior of the signal [18]. The same problem is encountered in the analysis of Rayleigh light scattering data; this is discussed in Chap 9 of Part 1 of this treatise.

The data acquisition system of Fig. 6 and 7 can of course be used in pulsed NMR (or NQR) with no hardware modifications, since the electrical signal to be digitized can be derived from any source. We have used the minicomputer system of Fig. 7, with minor interface additions and appropriate programs, with a pulsed NQR spectrometer in this laboratory, to collect spin-echo NQR data and to do Fournier transforms of coadded free induction decay transients.

Recently we also wrote programs to compute a continusou distribution of relaxation times from pulsed birefringence decay signals, by doing the inverse Laplace transform to derive coefficients in a polynomial fitting function, a procedure pioneered by Watanabe and Yoshioka [7,34].

Developments in on-line computerized electro-optic apparatus have been reported recently from several other laboratories [35-39].

ACKNOWLEDGMENTS

This research has been supported in part by the National Cancer Institute under research grants CA 12, 540-01 to -05 and by the Office of Naval Research. We thank Marcos F. Maestre and Colin F. MacKay for helpful suggestions leading to improvements in the manuscript.

REFERENCES

1. C. T. O'Konski and A. J. Haltner, J. Am. Chem. Soc., 78, 3604 (1956).

2. J. Schweitzer and B. R. Jennings, Biopolymers, 11, 1077 (1972).

3. C. M. Paulson, Ph.D. Thesis, University of California (1965). (Private communication, 1972).

4. S. Kobayashi, Biopolymers, 6, 1491 (1968).

5. C. T. O'Konski, K. Yoshioka, and W. J. Orttung, J. Phys. Chem., 63, 1558 (1959).

6. M. Matsumoto, H. Watanabe, and K. Yoshioka, Biopolymers, <u>6</u>, 929 (1968).

7. M. Matsumoto, H. Watanabe, and K. Yoshioka, Kolloid-Z. Z. Polymere, <u>250</u>, 298 (1972).

8. R. E. Smith and M. Morales, Bull. Math. Biophysics, <u>6</u>, 133 (1944).

9. W. Perl, Int. J. Appl. Rad. Isotopes, <u>8</u>, 211 (1960).

10. S. Krause and C. T. O'Konski, J. Am. Chem. Soc., <u>81</u>, 5082 (1959).

11. B. L. Brown, and B. R. Jennings, Sci. Instr., <u>3</u>, 195 (1970).

12. H. Benoit, Ann. Phys., <u>6</u>, 561 (1951).

13. M. Johnsen, Chem. Scripta, <u>1</u>, 149 (1971).

14. D. L. Kreider, R. G. Kuller, D. R. Ostberg, and F. W. Perkins, An Introduction to Linear Analysis, Addison-Wesley, Reading, Mass., 1966.

15. J. Mathews, and R. L. Walker, Mathematical Methods of Physics, Benjamin, New York, 1965.

16. C. Lanczos, Applied Analysis, Prentice-Hall, Englewood-Cliffs, N. J., 1956, p. 272.

17. R. A. Buckingham, Numerical Methods, Pitman, New York, 1957, p. 329-333.

18. F. A. Willers, Practical Analysis (R. T. Beyer, transl.), Dover, New York, 1948, p. 355-363 (Ger. Ed. 1928).

19. I. Isenberg and R. D. Dyson, Biophys. J., <u>9</u>, 1336 (1969).

20. D. J. Wilde, Optimum Seeking Methods, Prentice-Hall, Englewood-Cliffs, N. J., 1964.

21. D. R. Powell and J. R. McDonald, Computer J., <u>15</u>, 148 (1972).

22. F. James, Proc. of the 1972 CERN Computing and Data Proc. School, Pertisau, Austria. 10-24 September, 1972. (CERN 72-21).

23. R. Hooke and T. A. Jeeves, J. Assoc. Comp. Mach., <u>8</u>, 212 (1961).

24. D. G. Gardner, J. C. Gardner, G. Laush, and W. W. Meinke, J. Chem. Phys., <u>31</u>, 978 (1959).

25. E. C. Titchmarsh, Introduction to the Theory of Fourier Integrals, Oxford University Press, New York (1937).

26. R. F. Paley and N. Wiener, Fourier Transforms in the Complex Domain, American Mathematical Society, 1934.

27. A. Papoulis, Q. Appl. Math., 14, 405 (1957).

28. J. H. Renken and Frank Biggs, J. Computational Phys., <u>9</u>, 318 (1972).

29. W. D. Gwinn, Private communications, 1973.

30. L. B. Cohen, B. Hille, R. D. Keynes, D. Landowne, and
    E. Rojas, J. Physiol., 218, 205 (1971).

31. M. Kwan, M. Maestre, C. T. O'Konski, and L. S. Shepard,
    Submitted to Biopolymers.

32. T. C. Farrar and E. D. Becker, Pulse and Fourier Transform
    NMR, Academic, New York, 1971.

33. R. T. Bell, Introductory Fourier Transform Spectroscopy.
    Academic, New York, 1972.

34. K. Tsuji, H. Watanabe, and K. Yoshioka, Adv. Mol. Relax-
    ation Processes, 6, 1-14 (1975).

35. E. Fredericq and C. Houssier, Electric Dichroism and Electric
    Birefringence, Clarendon, Oxford, 1973, p. 80.

36. C. Houssier, Laboratory Practice, October, 562 (1974).

37. M. Tricot and C. Houssier in Polyelectrolytes, Technomic,
    Westport, CT, 1976, p. 43.

38. M. Isles and B. R. Jennings, Brit. Polymer J., March, 34
    (1976).

39. B. R. Jennings and H. J. Coles, Proc. Roy. Soc. (London)
    A348, 525 (1976).

Chapter 16

USE OF ROTATIONAL DIFFUSION COEFFICIENTS
FOR MACROMOLECULAR DIMENSIONS AND SOLVATION

Phil G. Squire

Department of Biochemistry
Colorado State University
Fort Collins, Colorado

I.   INTRODUCTION

In characterizing the properties of macromolecules in
solution, the frictional coefficients corresponding to the processes of translation and rotation assume crucial roles. These
coefficients can be determined with a high degree of accuracy.
Baldwin [1] and O'Donnell et al. [2] have shown that, by careful

work, an accuracy of 0.2% is attainable in measurements of the
sedimentation coefficient. Gosting [3] reports an accuracy of
0.1% in measurements of the translational diffusion coefficient.
The translational frictional coefficient can be calculated either
from the sedimentation coefficient or the diffusion coefficient
with no loss of accuracy. Current methods for the determination
of rotational diffusion coefficients appear [4] to be capable of
an accuracy of 1-2%.

While the translational and rotational frictional coefficients can be accurately determined, their interpretation in terms
of the shape and volume of the hydrodynamic domain associated with
the molecule is a more difficult problem. First of all, it is
clear that the experimentally determined frictional coefficient,
either for translation or rotation is a product of two contributions, one from shape, the other from the volume of the hydrodynamic unit, and it is not possible to evaluate these two
contributions by the measurement of a single hydrodynamic parameter. Useful estimates of the degree of hydration as well as
molecular dimensions may be deduced by introducing reasonable
assumptions based on other experimental methods, but a more rigorous approach, introduced by Scheraga and Mandelkern [5], is to
combine the results of pairs of hydrodynamic equations, eliminating the contribution due to hydrodynamic volume and solving for
the contribution due to shape. After this is done, the volume of
the hydrodynamic unit may be readily determined.

When the shape and volume contributions to a frictional
coefficient have been separated, it is necessary to select a model
that best represents the hydrodynamic properties of the molecule.
Having done this, the dimensions of the model may be calculated.
Several models have been treated including ellipsoids of revolution [6,7], rigid rod [8,9], random coils [10], and chains of
intermediate flexibility [11]. Accumulated hydrodynamic data
suggest that most proteins are reasonably compact hydrated structures and yet multisubunit complexes may have structures that are

not compact. The assembly of subunits of bacterial membrane
ATPase [12] and glutamine synthetase [13] would be more closely
approximated by a doughnut than an ellipsoid of revolution.

In Chap. 3 (Vol. 1) the theoretical basis of the Kerr
electro-optic effect of solutions of rigid macromolecules is presented, the instrumentation required for its measurement is described, and some results on macromolecules are discussed.
Chapter 4 provides a unified discussion of rotational theory.
Chapters 17, 18, and 19 contain extensive reviews of electro-optic
measurements on proteins, nucleic acids, and polyelectrolytes. In
this chapter, we briefly review available methods for determining
rotational and translational diffusion coefficients, present the
shape-dependent functions derived from combinations of theoretical
equations, and report the results of studies in which these functions have been utilized.

II.   DEFINITION OF ROTATIONAL DIFFUSION COEFFICIENTS

The theoretical parameters of the rotational diffusion
process may be easily visualized by considering the birefringence
relaxation experiment (Chap. 3, Sec. III). If the macromolecules
are initially aligned with the a axis in the direction of the
applied field, then after termination of the field the system
tends toward random orientation by rotational motion around the b
and c axes. After a lapse of time, t, following termination
of the orienting field, the disorder of the system may be
characterized in terms of two angles, ϕ_b and ϕ_c, the angular
displacements of the a axis resulting from rotation around the
b and c axes, respectively. (A similar experiment in which the
macromolecules were aligned with the b or c axes in the
direction of the orienting field would define the corresponding
parameter ϕ_a.) The rotational diffusion coefficient θ_i for
rotation around the i axis, where i = a, b, or c, is

defined in terms of $\overline{\phi_1^2}$ the mean value of the angular displacements squared, at time t by an equation due to Einstein (see Edsall, Ref. 14):

$$\theta_i = \frac{1}{2}\frac{\overline{\phi_i^2}}{t} \tag{1}$$

The rotational frictional coefficients ζ_i for rotation around these axes are defined in terms of the corresponding rotational diffusion coefficients θ_i by the equation

$$\theta_i = \frac{kT}{\zeta_i} \tag{2}$$

where k is the Boltzmann constant, and T the absolute temperature.

The dielectric dispersion relaxation time τ_a corresponds to rotation of the a axis around the b and c axes.

By symmetry, there are three dielectric relaxation times each corresponding to rotation of one of these axes around the other two. They are related to the rotational diffusion coefficients by the relationships,

$$\tau_a = \frac{1}{\theta_b + \theta_c}, \quad \tau_b = \frac{1}{\theta_a + \theta_c}, \quad \tau_c = \frac{1}{\theta_a + \theta_b} \tag{3}$$

If the macromolecule may be described as an ellipsoid of revolution, then the b and c axes are equal and

$$\tau_a = \frac{1}{2\theta_b}, \quad \tau_b = \tau_c = \frac{1}{\theta_a + \theta_b} \tag{4}$$

and, of course, if the molecule is spherical

$$\tau = \frac{1}{2\theta} \tag{5}$$

As pointed out in Chap. 3, the birefringence decay for rotation of the a axis of an ellipsoid of revolution ($a \neq b = c$) around the b and c axes is given by:

$$\frac{\Delta n}{n_0} = e^{-t/\tau_n} \tag{6}$$

where τ_n is the birefringence relaxation time and is related to θ_b by $\tau_n = 1/6\theta_b$. Since the dielectric dispersion relaxation times, τ_ε for an ellipsoid of revolution are given by Eq. (4) it follows that:

$$3\tau_n = \tau_\varepsilon = \frac{1}{2\theta_b} \tag{7}$$

Fluorescence depolarization experiments lead to the calculation of the harmonic mean relaxation time, τ_h, which is defined as follows:

$$\frac{1}{\tau_h} = \frac{1}{3}\left(\frac{1}{\tau_a} + \frac{1}{\tau_b} + \frac{1}{\tau_c}\right) \tag{8}$$

τ_h can also be expressed in terms of rotational diffusion coefficients by substitution of Eqs. (3) into (8). For an ellipsoid of revolution $\tau_b = \tau_c$, so

$$\frac{1}{\tau_h} = \frac{1}{3}\left(\frac{1}{\tau_a} + \frac{2}{\tau_b}\right) \tag{9}$$

The rotational frictional coefficient for spheres is

$$\zeta_0 = 8\pi\eta r^3 \tag{10}$$

where η is the viscosity of the solvent. Combining Eqs. (2), (5), and (10) we obtain the relaxation time of a sphere of radius r,

$$\tau_0 = \frac{4\pi\eta r^3}{kT} \tag{11}$$

In the treatment of data obtained from measurements of macromolecules that are hydrated and not spherical, it is convenient to separate the contributions due to the unhydrated mass of the molecule, its hydration, and its shape. For this purpose an identity due originally to Oncley (see Ref. 14) is convenient.

The rotational frictional coefficient for a given rotational mode is expressed in terms of two hypothetical frictional coefficients: ζ_e, the rotational frictional coefficient of a sphere of the same volume as the hydrated macromolecule, and ζ_0, that of a sphere of the volume the molecule would occupy if it were not hydrated.

Then the identity

$$\frac{\zeta}{\zeta_0} = \frac{\zeta}{\zeta_e} \frac{\zeta_e}{\zeta_0} \tag{12}$$

expresses the frictional ratio ζ/ζ_0 as the product of two ratios, the first representing the contribution due to shape, and the second, the contribution due to hydration. Even though it may not be possible to evaluate the frictional coefficients corresponding to each of the rotational modes, substitution of Eq. (12) into equations relating experimental quantities to frictional coefficients provides useful equations for the interpretation of the data in terms of shape and hydration. Several examples of this approach are provided later in this chapter.

In concluding this section, I acknowledge a debt to Professor Edsall for an earlier lucid treatment of this subject [14]. For a more extensive discussion, and references to the original literature, the reader is referred to Professor Edsall's chapter [14].

III.   MEASUREMENT AND INTERPRETATION OF THE ROTATIONAL DIFFUSION COEFFICIENT

A.   Fluorescence Depolarization

Among the methods available for the determination of rotational diffusion coefficients fluorescence depolarization appears to be the most versatile with respect to lack of restrictions on solvent composition. In particular, the high conductivities of aqueous solvents containing electrolytes at physiological and even higher concentrations introduce no difficulties.

Two distinctly different experimental methods are now in use, the steady state depolarization method and the pulse method. In the steady state depolarization method, only the harmonic mean of the three rotational relaxation times can be determined. In addition, it is essential that the experimental material contain a fluorescent chromophore having a fluorescence lifetime reasonably close to the harmonic mean relaxation time of the molecule under study. Very briefly, the steady state fluorescence depolarization experiment is conducted as follows: The sample is continuously irradiated with polarized light traveling in the y direction. The electric vector is aligned vertically, the z direction, and the fluorescence intensity is measured in the direction normal to both of these axes, the x direction. The fluorescence intensity is measured through an analyzer oriented vertically to give $I_{\parallel}$ and horizontally to give $I_{\perp}$. The polarization, p, is defined [16] as

$$p = \frac{I_{\parallel} - I_{\perp}}{I_{\parallel} + I_{\perp}} \tag{13}$$

The polarization is related to the harmonic mean relaxation time, τ_h, by the Perrin equation. Written for a polarized incident beam this equation (16) assumes the form

$$\frac{1}{p} - \frac{1}{3} = \left(\frac{1}{p_0} - \frac{1}{3}\right)\left(1 + \frac{3\tau^*}{\tau_h}\right) \tag{14}$$

Here p_0 is the intrinsic polarization, the polarization that would be observed in the absence of rotational processes occurring between absorption and emission, and τ^* is the lifetime of the excited state, a quantity which must be determined independently.

The molecule is provisionally considered spherical. By substitution of Eq. (11) and the relation $V_e = 4\pi r^3/3$, Eq. (14) assumes the form

$$\frac{1}{p} - \frac{1}{3} = \left(\frac{1}{p_0} - \frac{1}{3}\right)\left(1 + \frac{kT\tau^*}{\eta V_e}\right) \tag{15}$$

where k is the Boltzmann constant, T the absolute temperature, η the viscosity of the solvent, and V_e the volume of the equivalent hydrated sphere, or simply the volume of the hydrodynamic domain associated with the macromolecule in solution.

One then measures the polarization at different temperatures, and in solutions in which the viscosity is varied by adding a third component, usually glycerol, to slow down the relaxation process. Then by plotting $1/p$ vs. T/η, one obtains $1/p_0$ from the intercept and $\frac{1}{p_0} - \frac{1}{3} \frac{k\tau^*}{V_e}$ from the slope. The harmonic mean relaxation time may then be determined from V_e by means of the equation

$$\tau_h = \frac{3\eta V_e}{kT} \tag{16}$$

If one wishes to calculate τ_h for the standard state, water at $25^{\circ}C$, this is done at this point merely by evaluating η and T for this state.

Analogous to Eq. (12) we can write

$$\frac{\tau_h}{\tau_0} = \frac{\tau_h}{\tau_e} \frac{\tau_e}{\tau_0} \tag{17}$$

We recall two equivalent expressions for the volume of the unhydrated molecule assuming spherical shape.

$$V = \frac{4\pi r^3}{3} = \frac{M\bar{v}}{N} \tag{18}$$

Solving for r^3 and substituting into Eq. (11) we have

$$\tau_0 = \frac{3\eta M\bar{v}}{RT} \tag{19}$$

where τ_0 is the relaxation time the molecule would have if spherical and unhydrated. In interpreting τ_h/τ_0 in terms of shape and hydration we encounter problems common to the interpretation of all hydrodynamic measurements yielding a single experimental parameter. Since this is the first time we encounter this

problem in this chapter, we will describe it in reasonable detail
here.

1.   How may we determine τ_e/τ_0, the contribution due to
hydration?   The results of experiments by other methods must be
utilized in answering this question.   One can assume a "reasonable"
hydration value for the "typical" protein.   Edsall convincingly
argues (Ref. 14, p. 690) that soluble proteins are hydrated to the
extent of at least 10%, more frequently around 30% and in some
cases considerably higher values may be plausible.   If a value for
w,   the degree of hydration is assumed, τ_e/τ_0 may be readily
calculated,

$$\frac{\tau_e}{\tau_0} = \frac{V_e}{V_0} = \frac{\bar{v} + w}{\bar{v}} \tag{20}$$

Assuming a degree of hydration, $w = 0.2$ g H_2O/g   protein as is
often done, and a partial specific volume $\bar{v} = 0.73$ which is a
reasonable approximation for simple proteins,

$$\frac{\tau_e}{\tau_0} = \frac{0.93}{0.73} = 1.27 \tag{21}$$

For a discussion of hydration in the context of hydrodynamic
measurements, the reader is referred to Tanford [15].

2.   Having estimated τ_e/τ_0,   how then do we interpret τ_h/τ_e
in terms of shape?   This must be done in terms of a model.   The
decision concerning the appropriate model must also be based upon
external evidence.   Protein chemists usually assume that the shape
of the protein in solution may be approximated by an ellipsoid of
revolution.   The quantity τ_h/τ_e for prolate and oblate ellipsoids of revolution is plotted as a function of axial ratio in
Fig. 1.   It is clear that additional evidence would be required to
distinguish between the prolate and oblate model for $\tau_h/\tau_e < 2.2$,
but higher values would dictate the oblate model.

In the above discussion, I have implicitly excluded the
possibility of independent rotation of the fluorescent chroma-
phore.   For other applications of the fluorescence polarization

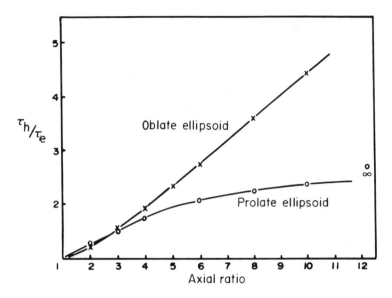

FIG. 1.   Ratio of the harmonic mean relaxation time, τ_h, of prolate and oblate ellipsoids of revolution to that of a sphere of equal volume, τ_e.   (Reproduced from Ref. 16 with permission of the author and publisher.)

method, the reader is referred to several reviews [16-20].

The nanosecond fluorescence depolarization method is a more recent development.   Developments in this field are reviewed by Tao [21] and by Cantor and Tao [22].   The theory is reexamined by Belford et al. [23].   In this method the sample is irradiated with brief pulses of light and the polarization is measured at selected time intervals after irradiation.   In this way, the time dependence of the polarization can be determined.   This method has two distinct advantages over the conventional method.   An independent measure of τ^* is unnecessary and in the case of distinctly nonspherical macromolecules, it is possible to calculate relaxation times for the individual rotational modes.   The advantages of knowing the two rotational relaxation times of an ellipsoid of revolution are illustrated in the following section.

B.   Dielectric Dispersion

The usefulness of dielectric dispersion measurements for
determining the size and shape of macromolecules was demonstrated
by Oncley and collaborators during the period 1935-1945.   In spite
of the early triumphs recorded by these workers, the method has
not been widely used.   This neglect appears to be due in part to
doubts raised by Kirkwood and Schumaker [24] and others [24a] as
to whether the observed dielectric increment was actually due to
rotation of a dipolar macromolecule.   This controversy is reviewed
by Takashima [25].   Subsequent studies [25,26] of proteins have
tended to reaffirm Oncley's original interpretation as applied to
these macromolecules.

Another deterrent was the fact that as a consequence of problems due to electrode polarization, the method is limited to solutions of low conductivity.   This limitation not only prevents one
from making measurements under physiological conditions, but even
more seriously, restricts measurements to the region of pH and
ionic strength at which proteins have minimum solubility.   Nevertheless, it seems likely that many additional biological
macromolecules which have been isolated in a state of high purity
have solubility properties that would permit their characterization by this method.

The details of dielectric dispersion measurements have been
reviewed elsewhere [25,27,28].   Here we merely indicate the nature
of the experimental data and outline an approach to its interpretation in terms of shape and hydration.

In these studies one measures the capacity of a condenser
filled with the material under study and does so as a function of
the frequency, ν, of a sinusoidal voltage.   The dielectric constant, ε' is related to the capacity of the condenser, C,   by

$$\varepsilon' = \frac{C}{C_0} \tag{22}$$

where C_0 is the capacity in vacuo. In measurements of
macromolecules in solution we are interested in the change in the
dielectric constant of the solution due to the addition of solute.
In addition, it is necessary to correct for electrode polarization
effects. We accomplish these purposes by defining the parameter,
the dielectric increment $\Delta\varepsilon'$,

$$\Delta\varepsilon' = \varepsilon'_p - \varepsilon'_s = \frac{C_p}{C_0} - \frac{C_s}{C_0} \tag{23}$$

Here C_s refers to the capacity of a salt solution, usually KCl,
having the same conductivity as the solution under study and C_p
that of the protein solution. Since the increment is dependent
upon protein concentration, it is also convenient to reduce this
parameter to unit concentration,

$$\frac{\Delta\varepsilon'}{c} = \frac{1}{c} \frac{C_p - C_s}{C_0} \tag{24}$$

For simplicity we assume that the dielectric increment results
from permanent dipole orientation, and assume a state of mass and
structural homogeneity of the sample. At low frequencies the dipolar macromolecules are able to reorient with the oscillating
field giving rise to an increase in the dielectric constant $\Delta\varepsilon'_0$.
At sufficiently high frequencies, the frequency greatly exceeds
the rate of rotation and the contribution of the macromolecule to
the dielectric constant, $\Delta\varepsilon'_\infty$ approximates that of a nonpolar
substance. The frequency corresponding to the midpoint of the
dispersion curve is known as the critical frequency ν_c.

    Theoretical analysis of the dispersion curve is facilitated
by normalizing the data. The normalized form is related to a
single relaxation time by an equation due to Debye

$$\frac{\Delta\varepsilon'_f - \Delta\varepsilon'_\infty}{\Delta\varepsilon'_0 - \Delta\varepsilon'_\infty} = \frac{1}{1 + (\omega\tau)^2} \tag{25}$$

where $\omega = 2\pi\nu$, and $\Delta\varepsilon'_f$ is the dielectric increment at a
frequency ν within the dispersion curve.

A single relaxation time would be expected in two situations, if the macromolecule were spherical or if the direction of the dipole corresponded to that of the major axis of an ellipsoid of revolution. If the macromolecule is an elongated ellipsoid of revolution and if the direction of the dipole moment intersects the major axis with the angle θ, then the molecule can follow the oscillating field by rotation either around the major or minor axes. This gives rise to two relaxation times τ_1 and τ_2. The two-term Debye equation has the form

$$\frac{\Delta\epsilon'_f - \Delta\epsilon'_\infty}{\Delta\epsilon'_0 - \Delta\epsilon'_\infty} = \frac{A_1}{1 + (\omega\tau_1)^2} + \frac{A_2}{1 + (\omega\tau_2)^2} \tag{26}$$

Here A_1 and A_2 denote the fraction of the total increment corresponding to the relaxation times τ_1 and τ_2. The angle θ is simply related to A_1 and A_2 by equation

$$\tan \theta = \sqrt{\frac{A_1}{A_2}} \tag{27}$$

Theoretical normalized dispersion curves for prolate ellipsoids of revolution with a dipole angle of 45° have been calculated by Oncley, and are reproduced in Fig. 2. The curve corresponding to axial ratio $= 1$ of course represents the single term Debye equation. Even in the most favorable case, $\theta = 45^\circ$, resolution of the two relaxation times is poor for axial ratios less than 9. In our study of bovine serum albumin [26], axial ratio 3.5, we become convinced that computer analysis of data of high precision was required in order to determine τ_1, τ_2, A_1, and A_2 with even modest accuracy. If one of the relaxation times can be accurately determined by an independent measurement, then the second can be determined from the dielectric dispersion curve. To this end, we [26] evaluated the τ_1 by birefringence relaxation and τ_2 from the dielectric dispersion curve using Eq. (7).

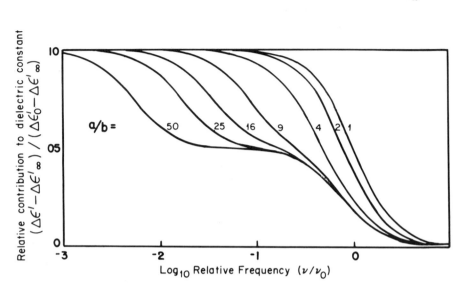

FIG. 2.   Dielectric dispersion curves for elongated ellipsoids
of revolution (according to Perrin) with $A_1/A_2 = 1$ and $\theta = 45°$.
Ordinate corresponds to left-hand side of Eq. (26). (Reproduced
from Ref. 63 with permission of the author and publisher.)

Having two relaxation times we are in a better position to
evaluate shape and hydration than by having only the harmonic
mean.

From the equations of Perrin, Wyman and Ingalls [29] have
related the ratio of τ_1/τ_2 to the axial ratio of oblate and prolate ellipsoids of revolution. Their graphical presentation of
this relationship is reproduced in Fig. 3. Here it may be seen
that τ_1/τ_2 varies little with axial ratio for oblate ellipsoids
and that if this ratio differs from unity by more than 10%, the
oblate model is excluded. For bovine serum albumin we [26]
determined $\tau_1 = 0.23$ μsec, $\tau_2 = 0.11$ μsec. The ratio
$\tau_1/\tau_2 = 2.09$ corresponds to a prolate ellipsoid with an axial
ratio equal to 3.0, the reciprocal of ρ, recorded in Fig. 3.

Continuing with bovine serum albumin as an example for the
calculation of molecular dimensions, we require the following
values: M = 66,700 Daltons, $\bar{v}$, the partial specific

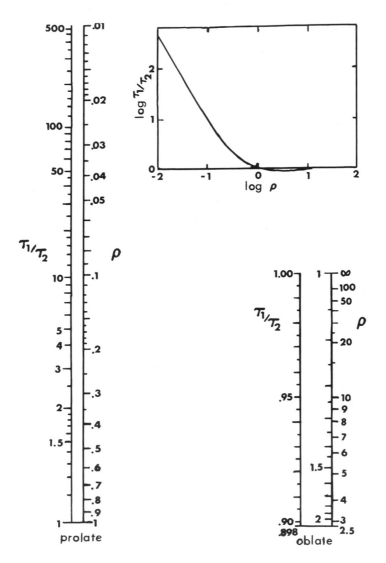

FIG. 3.   Ratio of the two relaxation times τ_1/τ_2 as a
function of axial ratio (ρ) for oblate and prolate ellipsoids
calculated from the equations given by Perrin.   (Reproduced from
Ref. 29 with permission of the author and publisher.)

Table 1

Axial Ratio of Hydrated Bovine Serum Albumin[a]

Data	Method of analysis	a/b
τ_{ε_1}, τ_{ε_2}, s, M(1 - $\bar{v}\rho$)	γ Function	3.0 [30]
τ_n, s, M(1 - $\bar{v}\rho$)	γ Function	3.6 [30]
τ_n, [η], M	δ Function	4.5 [30]
$\tau_{\varepsilon 1}$, [η], M	δ Function	3.5 [30]
s, [η], M	β Function	1 [5]
(τ_{ε_1} = 3τ_n), τ_{ε_2}	Perrin equations	3.0 [26]
s, [η], M	Λt^3	1 [47]
X-ray scattering		2.5 [48]
		3.9 [47]
K_s and [η]		3.2 [46]
Electron microscopy		3.5 [49]

[a]Reprinted from Biochemistry 7, 4268 (1968), copyright by
American Chemical Society, reprinted by permission.

volume = 0.734, and $\theta_b = 1/2\ \tau_1 = 2.18 \times 10^6\ \text{sec}^{-1}$. The rotational diffusion coefficient of the equivalent unhydrated sphere θ_0 can be calculated from these values, i.e.,

$$\theta_0 = \frac{RT}{6\eta M\bar{v}} = 9.51 \times 10^6\ \text{sec}^{-1} \tag{28}$$

Since $\zeta_b/\zeta_0 = \theta_0/\theta_b$, we can now calculate $\zeta_b/\zeta_0 = 9.51/2.18 = 4.37$. The ratio $\zeta_b/\zeta_e = \theta_e/\theta_b = 2.34$ for an axial ratio of 3.0 (Table 1, Ref. 30). From the ratio $\zeta_e/\zeta_0 = 4.37/2.34 = 1.86 = (\bar{v} + \bar{w})/\bar{v}$ we calculate $w = 0.63$ g/protein.

The dimensions of the hydrated molecule may be calculated from

$$\theta_e = \frac{\theta_0}{1.86} = \frac{9.51 \times 10^6}{1.86} = \frac{kT}{8\pi\eta ab^2} \tag{29}$$

where a is the axis of rotation and c is equal to b. Solving for a, we obtain a = 69 Å, and b = 23 Å, the values reported in Table III, Ref. 26.

Inherent in this approach is the assumption that the macromolecular sample is homogeneous with respect to mass and conformation. This assumption is of crucial importance and evidence for its validity is rarely supplied. Analyses of the sample by polyacrylamide gel electrophoresis would accomplish this purpose and would be an admirable general practice. Proteins often associate in solution. Even in highly purified preparations irreversible association products are frequently present. Gel filtration through an appropriate medium will usually remove irreversible association products as well as providing an estimate of the size distribution of the original sample. Gel filtration studies of highly purified serum albumin preparations [30,31] have shown that more than 10% of the protein may be irreversibly associated. While these remarks pertain to all the measurements discussed in this chapter, their applicability to dielectric dispersion measurements is perhaps most easily seen. In a study of

completely spherical molecules, the presence of 10-15% dimer might
give sufficient spreading to the curve that one would calculate
two critical frequencies. If heterogeneity were not recognized,
calculation of the parameters of a prolate ellipsoid of revolution
would probably result. Proof of conformational homogeneity is
unfortunately elusive.

              C.   Electric Birefringence Relaxation

     Since Chaps. 1 and 3 of this volume, as well as recent
reviews [32,35] are devoted to the theory of electric birefringence and the interpretation of results obtained by this method,
the discussion presented here is rather brief. We confine ourselves to the estimation of the parameters of shape and hydration
from birefringence relaxation data and define the parameters that
are useful in combination with hydrodynamic data obtained by other
methods.

     The birefringence relaxation method is an extremely powerful
one for this purpose since it provides a measure of the rotational
diffusion coefficient of the major axis of an asymmetric molecule,
around the two orthogonal minor axes. Among the measurable
hydrodynamic parameters, this one is the most sensitive to shape.
In contrast, most of the shape-sensitivity of rotation around the
individual axes is lost in forming the harmonic mean of the three
relaxation times and is also lost in averaging the translational
process over three translational modes.

     As pointed out in Chaps. 1 and 3, and evident from Eq. (6),
one can conveniently obtain the birefringence relaxation time of
a homogeneous axisymmetric macromolecule preparation from a plot
of log $\Delta\eta$ against time. As an example, O'Konski and Haltner
[36] studied a homogeneous preparation of tobacco mosaic virus in
dilute solution which gave a linear plot over a 25-fold range of
$\Delta\eta$ from which a relaxation time of 0.50 ± 0.02 msec was
calculated.

If the birefringence decay cannot be interpreted in terms of
a single relaxation time, this may be due to molecular inhomogene-
ity of the sample. Molecular association or degradation is frequently invoked. In such cases, an independent measure of the
molecular homogeneity of the sample actually used in the experiment would be highly desirable, because an alternative explanation
does exist. If the birefringence relaxation results from rotation
around the b and c axes and if $\theta_b \neq \theta_c$, this may also result
in two resolvable relaxation times. As an example, Pytkowicz and
O'Konski [37] found that interpretation of the birefringence
relaxation of snail hemocyanin required two relaxation times.
They interpreted their data by assuming that some of the protein
was in the form of a dimer. Ridgeway [38], while not disputing
the original interpretation, showed that the same data could be
interpreted in terms of an asymmetric ellipsoid of axial ratio
1:4.89:18.5. (It seems likely that the original interpretation
was the correct one. It has been known since the early work of
Brohult [33] that hemocyanins form association products with slow
rates of equilibrium. Furthermore, electron microscopy studies
by Fernandez Moran et al. [34] show that molluscan hemocyanins
indeed form cylindrical quaternary structures.)

In the event that the macromolecule is essentially axisymmetric and the birefringence axis is aligned with the symmetry
axis, and if the sample is monodisperse, then a single relaxation
time should be observed corresponding to rotation of the long axis
around the equal short axes. Even so, a single relaxation time
does not provide adequate information to characterize the molecule, since a single relaxation time can be interpreted in terms
of a range of values which may be assumed by the two parameters
--effective volume and axial ratio. Information from other
physical measurements must be utilized in the calculation of
molecular dimensions. Thus in their interpretation of the
birefringence relaxation of tobacco mosaic virus, O'Konski and

Haltner [36] utilized a value of 150 Å, for the diameter of the
hydrated virus. This value was taken from the x-ray diffraction
studies of Bernal and Fankuchen [39]. They were then able to
calculate the length of the rod from the rotational diffusion
coefficient by means of the equation of Burgers [8,40]

$$\theta = \frac{3kT(-0.80 + \ln 2a/b)}{8\pi\eta a^3} \tag{33}$$

where   2a   is the length of the rod,   2b   its diameter, and η
the viscosity of the medium.

There was some uncertainty in the length of the rod depending
upon the amount of hydration assumed, the best estimate being
3416 ± 15 Å, a value significantly greater than 2980 ± 10 Å
from electron microscopy. After considering and rejecting several
possible sources of error, they questioned the adequacy of the
Burgers equation. This led Broersma [9] to recalculate the frictional coefficient of a rotating cylinder and to derive an improved equation for the rotational diffusion coefficient,

$$\theta = \frac{3kT}{8\pi\eta a^3} \left[\ln 2p - 1.57 + 7\left(\frac{1}{\ln 2p} - 0.28\right)^2 \right] \tag{34}$$

where   p   is the ratio of rod length to diameter. Recalculation
of the length of the virus by means of this equation gave a value
of 3020 Å, in good agreement with the value 2980 Å from electron
microscopy.

Combination of the birefringence relaxation time with data
from dielectric dispersion measurements was described in the previous section of this chapter.

D.   Flow Birefringence

Prior to the application of the electric birefringence method
to the study of macromolecules, studies of flow birefringence had
already provided unequivocal evidence that tobacco mosaic virus

and fibrinogen, in particular, were highly extended rod-shaped
macromolecules. The essential features of the experimental method
and the results of early studies by this method are reviewed by
Cerf and Scheraga [41] and Tanford [15]. In this method birefringence results from the alignment of highly asymmetric macromolecules by a velocity gradient produced in the annular space between
two concentric cylinders, one rotating and one stationary. If
birefringence is observed, one has unequivocal evidence that the
macromolecule is not spherical since spheres would not be aligned
in a velocity gradient. Provided that the molecular weight is
known, it is possible to distinguish between rod-shaped and disc-
shaped macromolecules, since much higher velocity gradients are
required to align disc-shaped macromolecules. For rod-shaped
macromolecules, the calculated rotational relaxation time corresponds to rotation of the long axis around the two equal short
axes. Interpretation of this relaxation time in terms of shape
and hydration has been outlined earlier in this chapter. Since
this method is most useful in the study of highly asymmetric rod-
shaped macromolecules, frequent use is made of the limiting form
of the Perrin equation for axial ratios greater than 5,

$$\zeta = \frac{16\pi\eta a^3}{3[-1 + 2 \ln (2a/b)]} \tag{35}$$

where a and b are the long and short semiaxes, respectively.
This equation is exceptionally useful for measuring macromolecular
lengths, because the frictional coefficient, ζ, is highly dependent upon a, and relatively insensitive to b.

The early studies [42] of flow birefringence of serum albumin
solutions provided unequivocal evidence that the molecule was
nonspherical. Tanford's [15] analysis of these data along with
intrinsic viscosity data led him to the conclusion that the best
approximation to the actual shape of the macromolecule would be a
reasonably compact prolate ellipsoid of revolution of axial ratio
a/b $\simeq$ 4.

In studies of highly asymmetric macromolecules, flow
birefringence serves as a valuable complement to electric birefringence since the direction of alignment in a flow gradient is
easily deduced, while in electric birefringence it may be uncer-
tain. Examples of valuable conclusions based on data from both
measurements are given in Chapter 14.

IV.   METHODS OF COMBINING THE RESULTS
FROM VARIOUS HYDRODYNAMIC MEASUREMENTS

The difficulties in the interpretation of the rotational
frictional coefficient in terms of shape and hydration in the absence of data from other kinds of measurements has been frequently
referred to in this chapter. The same problems also occur in the
interpretation of the translational frictional coefficient and the
intrinsic viscosity. This difficulty arises largely from the fact
that the intrinsic viscosity $[\eta]$, the frictional ratio for
translation, f/f_0, and for rotation ζ/ζ_0 are all products of
two terms, one due to hydration, the other to nonspherical shape.
By combining the results of various hydrodynamic measurements,
however, it is possible to separate these two effects and, assuming an appropriate model, calculate the molecular dimensions and
degree of hydration.

A.   Graphical Methods

In the early work, treatment of combined hydrodynamic data
was accomplished by graphical methods. Wyman and Ingalls [29]
constructed several nomograms in which experimental quantities
were related to the axial ratio and degree of hydration of prolate
and oblate ellipsoids of revolution and illustrated their use with
hydrodynamic data obtained from studies of myoglobin and hemoglobin. Since these proteins had been studied by a rather large

number of experimental methods, the distribution of the total
frictional ratios into the shape-dependent and volume-dependent
parts was overdetermined. By successive approximations through
the alignment charts and by making reasonable allowance for
experimental error, it was quite convenient to arrive at a self-
consistent model that fit all the hydrodynamic data. Even so,
they were unable to determine whether the hemoglobin was best
described as a prolate or oblate ellipsoid of axial ratio 2.7.

Oncley accomplished the same purpose by means of nomograms in
which frictional ratios were plotted as contour diagrams as a
function of axial ratio and hydration. These contour diagrams
have been reproduced in various monographs (i.e., Refs. 14 and
43). If the results of several hydrodynamic measurements are
introduced into these contour diagrams, along with reasonable
estimates of experimental error, the area of overlap defines the
parameters of an equivalent ellipsoid of revolution that best
represents all the hydrodynamic properties of the molecule.
Several examples of this approach are given on page 562 of Ref.
43.

B. The Shape-Dependent Functions

Subsequent to the graphical solutions used by these early
workers, Scheraga and Mandelkern [5] and Sadron [44] introduced an
analytical approach that was subsequently extended by Scheraga
[45], Squire et al. [30], and Squire [4]. The approach is very
simple in concept. Recognizing that the frictional ratios both
for translation and rotation as well as the intrinsic viscosity
are all linear functions of the volume of the hydrated macromolecule, V_e as well as the shape, one can eliminate V_e between
pairs of equations thus obtaining shape functions that are
independent of volume. Since these functions may be calculated
from experimental data alone, they are true molecular constants
and are free of any assumptions regarding models, but if they are

to be interpreted in terms of molecular dimensions, a model must be assumed. Once the shape-dependent contribution is evaluated, the volume of the hydrated macromolecule, may be calculated from either of the three equations given in Table 2. Four types of shape functions have been derived in this way, the β and δ functions [5], the three γ functions [30] and the ψ function [4]. The Λt^3 function of Sadron [44] is equivalent to the β function, and the μ function [5] is equivalent to one of the three γ functions.

The β function is obtained by eliminating V_e from the pair of equations for the translational frictional coefficient and the intrinsic viscosity. Since the translational frictional coefficient can be calculated either from the diffusion coefficient, D, or the sedimentation coefficient, s, the β

Table 2

Effective Hydrodynamic Volume, V_e, and Hydration of BSA Axial Ratio 3.5[a]

Equations	V_e (cm^3)	Hydration (V_{H_2O}/V_e)
(1) $V_e = \dfrac{100 M [\eta]}{N \nu}$	10.97×10^{-20}	0.26
(2) $V_e = \dfrac{4\pi}{3}\left(\dfrac{f_e}{6\pi\eta_0}\right)^3$	12.10×10^{-20}	0.33
(3) $V_e = \dfrac{\zeta_e}{6\eta_0}$	10.48×10^{-20b} [30]	0.22^b
	12.26×10^{-20c} [30]	0.34^c
(4) x-ray scattering (pH 5.1)	13.0×10^{-20} [46]	

[a]Reprinted from Biochemistry 7, 4270 (1968), copyright by American Chemical Society, reprinted by permission.

[b]With θ_b obtained from dielectric dispersion.

[c]With θ_b obtained from birefringence decay.

function may be related to these experimental parameters by two
equivalent expressions,

$$\beta \equiv \frac{D[\eta]^{1/3}M^{1/3}\eta_0}{kT} = \alpha'F\nu^{1/3} \tag{36}$$

and

$$\beta \equiv \frac{Ns[\eta]^{1/3}\eta_0}{M^{2/3}(1 - \bar{v}\rho)} = \alpha'F\nu^{1/3} \tag{37}$$

where $[\eta]$ is the intrinsic viscosity, $\bar{v}$ the partial specific
volume, and M the molecular weight. Theoretical values of β
may also be calculated for ellipsoids of revolution of various
axial ratios from the expression $\alpha'F\nu^{1/3}$. Tables of the β
function are available for prolate and oblate ellipsoids of revolution over a wide range of axial ratios [5,45]. All experimental
quantities must of course be converted to the same standard state,
usually infinite dilution in water at 25°C or 20°C. η_0 and ρ
are the viscosity and density of water corresponding to the
standard state.

The β function is quite insensitive to axial ratio. As a
consequence, it is a valuable equation for the determination of
molecular weights from measurements of the sedimentation coefficient and the intrinsic viscosity.

The δ function is calculated from the intrinsic viscosity
and the rotational diffusion coefficient,

$$\delta = \frac{600}{Nk}\left(\frac{\eta_0\theta}{T}\right)[\eta]M = J\nu \tag{38}$$

Actually, separate δ functions would be required for each of the
rotational diffusion coefficients, but only the δ function
corresponding to θ_b for ellipsoids of revolution has been tabulated [5,45]. Due to the fact that this function is calculated
from the rotational diffusion coefficient of the long axis around

the short, it is much more sensitive to axial ratio than is the β function.

The γ functions are calculated from the rotational and translational diffusion coefficients. Since in the general case there are three rotational diffusion coefficients, there are also three γ functions. They are defined as follows [30]:

$$\gamma_a = \left(\frac{f}{f_e}\right)^3 \frac{\theta_b + \theta_c}{2\theta_c} \tag{39a}$$

$$\gamma_b = \left(\frac{f}{f_e}\right)^3 \frac{\theta_a + \theta_c}{2\theta_c} \tag{39b}$$

$$\gamma_c = \left(\frac{f}{f_e}\right)^3 \frac{\theta_a + \theta_b}{2\theta_c} \tag{39c}$$

For ellipsoids of revolution, $\theta_b = \theta_c$ and $\gamma_b = \gamma_c$. The functions γ_a and γ_b were tabulated by Squire et al. [30] and are presented graphically in Fig. 4. Here we see that the two γ functions are quite sensitive to shape for prolate ellipsoids of revolution and the values diverge with increasing axial ratio. For oblate ellipsoids of revolution γ_a is approximately equal to γ_b for all axial ratios. Thus, determination of the γ functions from experimental data provides one of our most useful methods for distinguishing between prolate and oblate ellipsoids of revolution.

Since the translational frictional coefficient may be determined either from s or D, experimental values for the γ functions may be calculated [30] from either parameter. If the sedimentation coefficient is used and if this value as well as the rotational diffusion coefficients are extrapolated to infinite dilution and corrected to 25° in water, the experimental value of the γ functions may be calculated from the equation

$$\gamma_i = 2.612 \times 10^{-57} \left[\frac{M(1 - \bar{v}\rho)}{s}\right]^3 \sum_{j \neq i} \theta_j \tag{40}$$

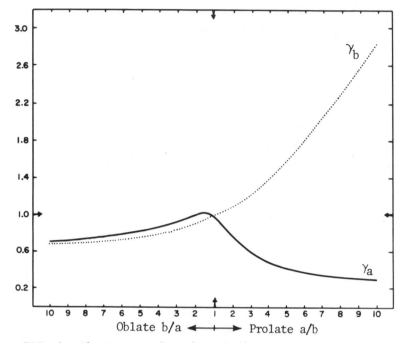

FIG. 4.   The two γ functions defined by Eqs. (39a) and (39b) for prolate and oblate ellipsoids of revolution.

The ψ function is calculated from the harmonic mean relaxation time and the sedimentation coefficient by

$$\frac{M(1 - \bar{\nu}\rho)}{s}\left(\frac{1}{\tau_h}\right)^{1/3} = A\psi \tag{41}$$

where

$$\psi = \left[\frac{(\theta_a + \theta_b + \theta_c)}{3\theta_e}\right]^{1/3} f/f_e \tag{42a}$$

and

$$A = 6\pi\eta_0 N\left(\frac{kT}{4\pi\eta_0}\right)^{1/3} \tag{42b}$$

If the experimental quantities on the left side of Eq. (41) are converted to 25°, $A = 7.243 \times 10^{18}$, and at 20°, $A = 7.492 \times 10^{18}$. The ψ function tabulated for prolate and oblate ellipsoids of revolution is also available [4].

The ψ function is calculated from data in which the experimental parameters are averaged over three rotational and three translational modes. As a consequence, it is relatively insensitive to axial ratio and is primarily useful [4] as an equation of consistency.

C.   Analysis of Data from Studies of Bovine Serum Albumin

Perhaps the most thorough study of the hydrodynamic properties of a single protein by the methods outlined here was an analysis of bovine serum albumin [30]. The preparations used in these experimental studies had been defatted and the oligomeric irreversible association products had been removed by gel filtration and the mass homogeneity was demonstrated by sedimentation velocity experiments. The β, δ, and γ functions were used in estimating the axial ratio of the prolate ellipsoid of revolution that best approximated the hydrodynamic properties of the protein, the prolate model having been excluded since γ_a, found to be 0.52 is well below the minimum value for any oblate ellipsoid of revolution (see Fig. 4). The results of this analysis are reproduced in Table 1. Also included in Table 1 are conclusions drawn from x-ray scattering data, as well as the application of the "rule of Wales and Van Holde." In this method the concentration dependence of the sedimentation coefficient K_s is combined with the intrinsic viscosity $[\eta]$ to yield the axial ratio assuming an ellipsoidal model. This rule was applied to globular proteins including bovine serum albumin by Creeth and Knight [46].

Yang [50] has pointed out that four groups have calculated essentially equal values of β for serum albumin of about 2.05×10^6 which is less than the minimum theoretical value of

2.12×10^6. Several [44,50-52] explanations have been advanced to explain this discrepancy, none of which is completely satisfactory. The Λt^3 function also leads to an inconsistent value. The δ functions, on the other hand, lead to the calculation of axial ratios that are unusually high. Thus all the shape functions that include the intrinsic viscosity appear to give incorrect values.

With the exception of values based on viscosity measurements, the axial ratios listed in Table 1 are in reasonable accord with the value 3.5. Taking this as our best value, we then calculate the hydrodynamic volume, V_e, using the equations given in Table 2 taken from Yang's review [50]. Here we see that four of the calculations lead to values for V_e ranging from $(10.8 - 13.0) \times 10^{-20}$ cm^3 while the value calculated from the intrinsic viscosity is somewhat lower. Hydration values calculated from the values of V_e are listed in the final column.

V.   DISCUSSION AND PERSPECTIVES

In addition to introducing an analytical method for separating the contributions due to molecular volume and asymmetry, Scheraga and Mandelkern [5] and subsequently Scheraga [45] have criticized the practice of expressing V_e as the sum of two terms

$$V_e = \frac{M\bar{v}}{N} + \frac{Mw}{N\rho_0} \tag{43}$$

where ρ_0 is the density of the solvent and w is the number of grams of solvent bound per gram of protein. They argue that this separation into two terms implies that the molecular domain consists of two parts, with $M\bar{v}/N$ the volume of the dry protein and $Mw/N\rho_0$ the volume of the bound water, and further implies that $\bar{v}$ is the reciprocal of the density of the dry protein. That the latter is not correct for all solutes is dramatically pointed out by the fact that $MgSO_4$ has a negative value of $\bar{v}$ at low

concentrations. Simple proteins, on the other hand, have partial
specific volumes of about 0.75 cm^3/g, a value in reasonable
agreement with the reciprocal of their dry densities. Specifically to this point, Bull [53] calls our attention to early work [54]
showing that the partial specific volume of egg albumin at pH 4.8
and 25° is 0.7461, whereas the specific volume of the dry protein
as measured by the volume of gaseous hydrogen displaced at 25° is
0.7407. Thus, while the use of Eq. (43) for calculation of hydration would lead to serious error when applied to some substances,
this is not likely to be true of proteins. (For a more extensive
discussion see Yang's review beginning on page 338 [50].) In our
calculations reported in Table 2, we have first calculated V_e by
methods not subject to these objections and subsequently calculated the degree of hydration in spite of them, simply because it
is a convenient parameter conceptually.

The structures of several proteins are now known from x-ray
crystallography. A supplement to a monograph by Dickerson and
Geis [55a] contains computer-generated stereoscopic drawings of
the three-dimensional structures of eight protein molecules.
These proteins provide valuable models for testing the hydrodynamic approach and for the further refinement of theory. We note
[55a] that hemoglobin appears to be roughly spherical in contrast
to earlier conclusions cited on page 587 that the molecule could
best be described as a prolate or oblate ellipsoid of axial ratio
2.7. Upon reading this manuscript, Heck [55b] pointed out that if
the results of more recent hydrodynamic measurements are used,
this discrepancy disappears. The values $s_{20,w} = 4.3 \times 10^{-13}$ sec
[55c], $\theta_{b,20,w} = 4.12 \times 10^6$ sec^{-1} [55d], M = 64,793 from the
amino acid composition plus four oxygen molecules and $\bar{v} = 0.749$
[55e] are sufficient for the calculation of γ_a. The constant in
Eq. (40) converted to 20° is 2.098×10^{-57}. The calculated value
for γ_a is 0.998, essentially the value for a sphere, Figure 4
[30].

Before attempting a closer comparison of data obtained by the two methods the accuracy of the hydrodynamic data requires examination. The value of θ_{20} used here was calculated from a published relaxation time measured at 17° with a precision of $\pm 1.5\%$. The data were converted to 20°, but they were not extrapolated to infinite dilution as required for our purposes.* The sedimentation coefficient was also determined at a single concentration, 0.7%, and no estimate of the precision of the measurement was made. It was clearly not the objective of the paper to determine s with maximum precision. The relative error in $s_{20,w}$ at infinite dilution is tripled in computing γ. The value of $\bar{v}$ determined in 1926 has survived the test of time, but it is important to point out that when γ is computed in this way, the relative error in $\bar{v}$ is multiplied by a factor of about 9. In contrast, by using the buoyancy-corrected molecular weight $M(1 - \bar{v}\rho)$, calculated directly from the sedimentation equilibrium experiment, one avoids this source of error.

In view of these considerations, the calculated value of γ might have a probable error of around 5%. This range of γ would accommodate a prolate axial ratio as high as 1.2, and an oblate axial ratio as high as 2.6. Thus a critical test of the hydrodynamic method based on the hemoglobin molecule requires a more careful determination of these hydrodynamic parameters, but at least they do not appear to be inconsistent.

The question of whether a protein has the same conformation in solution as in the wet crystals used for x-ray studies has been of considerable interest. From recent reviews [55f,55g] on this subject one can conclude that if structural differences exist, they are minor and would not be detected by hydrodynamic methods. For multisubunit enzymes, on the other hand, the question of whether these assemblies observed in electron micrographs have the

_____

*Since hemoglobin dissociates at high dilution, data for characterization of the tetrameric unit should be obtained in the concentration range where dissociation is negligible, and these data should be extrapolated to infinite dilution.

same quaternary structure as the assembly in solution is more
open. Examples of multisubunit enzymes, whose structures have
been determined by electron microscopy and other methods, would
include membrane ATPase and glutamine synthetase, mentioned in the
introduction to this chapter, as well as aspartate transcarbamy-
lase. Other examples are given by Klotz et al. [56]. The x-ray
structure of the latter enzyme has recently been published by
Lipscomb and his colleagues [57]. Since the subunits are held
together by noncovalent bonds, their possible rearrangement during
preparation for electron microscopy is a matter of serious concern.

One may calculate the expected rotational and translational
frictional coefficients of the assembly observed in the electron
microscope by means of the shell model calculations similar to
those of Bloomfield [58-61] and Hearst [62], and compare these
values with the measured ones. This was done, e.g., for the BSA
dimer [30]. A review of the theoretical problems associated with
this approach and an attempt at their resolution has been recently
published by Yamakawa [64] (see also Ref. 66).

A recent development in instrumentation that shows considerable promise for the measurement of the translational diffusion
coefficient of macromolecules in solution as well as the rotational diffusion coefficient of very large geometrically anisotropic
particles should be called to the attention of the reader. It is
basically a light scattering technique in which a laser is used as
the light source and line-broadening of the monochromatic light
source, due to the Doppler effect, provides the input data for
these calculations. A review of several applications of this
method has been published by Pecora [65].

It is clear that electro-optic studies of macromolecules in
solution have contributed greatly to our understanding of their
conformation, and have focused attention on a significant error in
at least one hydrodynamic equation leading to an improved form.
These contributions in the area of hydrodynamics are in addition
to the determination of electro-optic parameters that are uniquely

determinable by these methods and are described in detail in other chapters of this book. In this chapter, considerable emphasis has been placed on the inherent difficulties in attempts to accurately determine the parameters of shape and hydration, and the use of combined hydrodynamic data for this purpose. It is hoped that this emphasis has not detracted from a recognition of the remarkable achievements that have resulted from electro-optic studies that have already been made, or from a perception of the potential for important contributions in the future.

ACKNOWLEDGMENTS

This chapter was published with the approval of the Director of the Colorado State University Experiment Station as Scientific Series Paper No. 1803. The work was supported by National Science Foundation Grant GB 21425. Drs. Victor Bloomfield and Henry Heck read this chapter in manuscript, and their comments and suggestions were greatly appreciated.

REFERENCES

1. R. L. Baldwin, Biochem. J., $\underline{65}$, 503 (1957).

2. I. J. O'Donnell, R. L. Baldwin, and J. W. Williams, Biochim. Biophys. Acta, $\underline{28}$, 294 (1958).

3. L. J. Gosting, Adv. Prot. Chem., $\underline{11}$, 429 (1956).

4. P. G. Squire, Biochim. Biophys. Acta, $\underline{221}$, 425 (1970).

5. H. A. Scheraga and L. Mandelkern, J. Am. Chem. Soc., $\underline{75}$, 179 (1953).

6. R. Gans, Ann. Physik, $\underline{86}$, 628 (1928).

7. F. Perrin, J. Phys. Radium, $\underline{5}$, 497 (1934).

8. J. M. Burgers, 2nd Report on Viscosity and Plasticity, Amsterdam Acad. Sci., Nordman, Amsterdam, Chap. 3 (1938).

9. S. Broersma, J. Chem. Phys., $\underline{32}$, 1626 (1960).

10. B. H. Zimm, J. Chem. Phys., $\underline{24}$, 269 (1956).

11.   J. E. Hearst, J. Chem. Phys., 38, 1062 (1963).

12.   H. P. Schmebli, A. E. Vatter, and A. Abrams, J. Biol. Chem.,
      245, 1122 (1970).

13.   D. F. Deuel, A. Ginsburg, J. Yeh, E. Shelton, and
      E. R. Stadtman, J. Biol. Chem., 245, 5195 (1970).

14.   J. T. Edsall, in The Proteins (H. Neurath and K. Bailey,
      eds.), Vol. 1B, Academic, New York, 1953, Chap. 7.

15.   C. Tanford, Physical Chemistry of Macromolecules, Wiley, New
      York, 1961.

16.   G. Weber, Adv. Protein Chem., 8, 415 (1953).

17.   L. Brand and B. Witholt, Enzymol., 11, 776 (1967).

18.   S. Udenfreund, Fluorescence Assays in Biology and Medicine,
      Academic, New York, 1962, Chap. 6.

19.   R. F. Steiner and H. Edelhoch, Chem. Rev., 62, 457 (1962).

20.   A. J. Pesce, C. G. Rosen, and T. L. Pasby, Fluorescence
      Spectroscopy, An Introduction for Biology and Medicine,
      Marcel Dekker, New York, 1971.

21.   T. Tao, Biopolymers, 8, 609 (1969).

22.   C. R. Cantor and T. Tao, Proc. Nucleic Acid Res., 2, 31
      (1971).

23.   G. G. Belford, R. L. Belford, and G. Weber, Proc. Nat. Acad.
      Sci. U.S., 69, 1392 (1972).

24.   J. G. Kirkwood and J. B. Schumaker, Proc. Nat. Acad. Sci.
      U.S., 38, 855 (1952).

24a. C. T. O'Konski, J. Chem. Phys., 23, 1559 (1955); J. Phys.
      Chem., 64, 605 (1960).

25.   S. Takashima, in Physical Principles and Techniques of
      Protein Chemistry, Part A (S. J. Leach, ed.), Academic, New
      York, 1969, Chap. 6.

26.   P. Moser, P. G. Squire, and C. T. O'Konski, J. Phys. Chem.,
      70, 744 (1966).

27.   H. P. Schwan, Phys. Tech. Biol. Res., 6, 323 (1963).

28.   D. Rosen, in A Laboratory Manual of Analytical Methods of
      Protein Chemistry, Vol. 4 (P. Alexander and H. P. Lundgren,
      eds.), 1966, p. 191.

29.   J. Wyman, Jr. and E. N. Ingalls, J. Biol. Chem., 147; 297
      (1943).

30.   P. G. Squire, P. Moser, and C. T. O'Konski, Biochemistry, 7,
      4261 (1968).

31.   K. O. Pedersen, Arch. Biochem. Biophys. Suppl, 1, 157 (1962).

32. C. T. O'Konski, Encycl. Polymer Sci. Technol., 9, 551 (1968).

33. S. Brohult, Nova Acta Reg. Soc. Sci. Ups. [4], 12, No. 4
     (1940).

34. H. Fernandez Moran, E. F. J. van Bruggen, and M. Ohtsuki, J.
     Mol. Biol., 16, 191 (1966).

35. K. Yoshioka and H. Watanabe, in Physical Principles and
     Techniques of Protein Chemistry, Part A (S. J. Leach, ed.),
     Academic, New York, 1969, Chap. 7.

36. C. T. O'Konski and A. J. Haltner, J. Am. Chem. Soc., 78, 3604
     (1956).

37. R. M. Pytkowicz and C. T. O'Konski, Biochim. Biophys. Acta,
     36, 466 (1959).

38. D. Ridgeway, J. Am. Chem. Soc., 88, 1104 (1966).

39. J. D. Bernal and I. Fankuchen, J. Gen. Physiol., 25, 111, 149
     (1941).

40. J. M. Burgers, Verhandel. Konikl. Ned. Akad. Wetenschap.
     Afdel. Natuurk, Sec. I, 16, 113 (1938).

41. R. Cerf and H. A. Scheraga, Chem. Rev., 51, 185 (1952).

42. J. T. Edsall and J. F. Foster, J. Am. Chem. Soc., 70, 1860
     (1948).

43. E. J. Cohn and John T. Edsall, Proteins, Amino Acids and
     Peptides, Reinhold, New York, 1943.

44. C. Sadron, Prog. Biophys. Biophys. Chem., 3, 237 (1953).

45. H. A. Scheraga, Protein Structure, Academic, New York, 1961.

46. J. M. Creeth and C. K. Knight, Biochim. Biophys. Acta, 102,
     549 (1965).

47. M. Champaigne, J. Chim. Phys., 54, 378 (1957).

48. V. Luzatti, J. Witz, and A. Nicolaieff, J. Mol. Biol., 3, 379
     (1961).

49. A. Chatterjee and S. N. Chatterjee, J. Mol. Biol., 11, 432
     (1965).

50. J. T. Yang, Adv. Protein Chem., 16, 323 (1961).

51. C. Tanford and J. G. Buzzell, J. Phys. Chem., 60, 225 (1956).

52. H. K. Schachman, Ultracentrifugation in Biochemistry,
     Academic, New York, 1950.

53. H. B. Bull, An Introduction to Physical Biochemistry, Davis,
     Philadelphia, 1964, p. 339.

54. H. B. Bull, Cold Spring Harbor Symp., Quant. Biol., 6, 140
     (1938).

55a. R. E. Dickerson and J. Geis, The Structure and Action of Proteins, Harper and Row, New York, 1969.

55b. Henry Heck, Personal communication.

55c. F. J. Gutter, H. H. Sober, and E. A. Peterson, Arch. Biochem. Biophys., 62, 427 (1956).

55d. P. Schlecht, H. Vogel, and A. Mayer, Biopolymers, 6, 1717 (1968).

55e. T. Svedberg and R. Fahraeous, J. Am. Chem. Soc., 48, 430 (1926).

55f. J. A. Rupley, Structure and Stability of Biological Macromolecules (S. N. Timascheff and G. D. Fasman, eds.), Marcel Dekker, New York, 1969, p. 291.

55g. J. Drenth, Enzymes: Structure and Function, 29, 1 (1972). North Holland/American Elsevier.

56. I. M. Klotz, Neal R. Langermann, and Dennis W. Damall, Ann. Rev. Biochem., 39, 25 (1970).

57. S. G. Warren, B. E. P. Edwards, D. R. Evans, D. C. Wiley, and W. N. Lipscomb, Proc. Nat. Acad. Sci., 70, 1117 (1973).

58. V. A. Bloomfield, Biochemistry, 5, 684 (1966).

59. V. A. Bloomfield, W. O. Dalton, K. E. Van Holde, Biopolymers 5, 135 (1967).

60. D. P. Felson and V. A. Bloomfield, Biochemistry, 6, 1650 (1967).

61. R. J. Douthart and V. A. Bloomfield, Biochemistry, 8, 2225 (1969).

62. J. E. Hearst, J. Chem. Phys., 38, 1062 (1963).

63. J. L. Oncley, J. Phys. Chem., 44, 1103 (1940).

64. H. Yamakawa, J. Chem. Phys., 53, 436 (1971).

65. R. Pecora, Nature [Phys. Sci.], 231, 73 (1971).

66. H. Yamakawa and J. Yamaki, J. Chem. Phys., 58, 2049 (1973).

FURTHER READING

E. O. Forster and A. P. Minton, Phys. Meth. Macromol. Chem. 2, 185 (1972).

L. D. Kahn, Meth. Enzym., 26, 323 (1972). (Chapter 14, Electric Birefringence.)

M. Kasai and F. Oosawa, Meth. Enzymol., 26, 289 (1972). (Ch. 13, Flow Birefringence.)

S. Takashima, Meth. Enzym., 26, 337 (1972) (Chapter 15, Dielectric Dispersion).

J. Yguerabide, Meth. Enzym., 26, 498 (1972) (Chapter 24, Nanosecond Fluorescence Spectroscopy of Macromolecules.)

Chapter 17

ELECTRO-OPTICS OF POLYPEPTIDES AND PROTEINS

Koshiro Yoshioka

Department of Chemistry
College of General Education
University of Tokyo
Komaba, Meguroku, Tokyo, Japan

I. ELECTRO-OPTICS OF POLYPEPTIDES

A. Introduction

In this chapter a brief account will be given of the
application of electro-optic methods to studies of synthetic
polypeptides (polymerized amino acids) and proteins in solution.
Synthetic polypeptides have been extensively investigated by a
variety of physicochemical methods during the past couple of
decades, because they serve as simple models for proteins and,
moreover, exhibit unique properties, such as the helix-coil transition. It has been established that synthetic polypeptides can
assume ordered conformations, such as α helix or β structure,
even in the solution state [1]. In the α helix model of
Pauling et al. [2], the N-H and C=O groups (which are
intramolecularly hydrogen-bonded) are directed almost along the
helix axis. From the vectorially additive nature of the residue
dipole moment along the helix axis, a large resultant dipole moment is expected for the whole molecule. Such dipolar macromolecules will be easily aligned in solution under the action of an
applied electric field, leading to a variety of anisotropy effects.

Among the electro-optic effects hitherto studied on polypeptide solutions are ranked the electric birefringence, the electric
dichroism, the optical rotation and circular dichroism in an
electric field, the electric field light scattering, and the
electric-streaming birefringence. Measurements of these electro-
optic properties yielded information about the mechanism of electric orientation of polypeptide molecules in solution, the
permanent dipole moment, the anisotropy of electric polarizability,
the rotational diffusion constant and its distribution (the length
and its distribution in the case of rigid rodlike molecules), the
molecular flexibility, the anisotropy of optical polarizability,
the direction of the chromophoric transition moment, the optical
rotation and circular dichroism for light incident parallel and
perpendicular to the helix axis, the mode of molecular association,

the molecular cluster in liquid crystals, the helix-coil transition, and other conformation changes. The results obtained up to the end of 1973 will be reviewed in the following sections. Concerning the optical rotation and circular dichroism in an electric field of polypeptide solutions, the reader should refer to Chapter 10 (Vol. 1).

B. Electric Birefringence of Poly-γ-benzylglutamate

In 1956 Doty et al. [3] evidenced from light scattering and intrinsic viscosity studies that poly-γ-benzyl-L-glutamate (PBLG) exists as rigid, rodlike helices in certain solvents (i.e., chloroform, dichloroethane, dimethylformamide) and as random coils in dichloroacetic acid.

In the following year, Tinoco [4] studied the transient electric birefringence of two PBLG samples of different molecular weight $(M_W = 8.4 \times 10^4$ and $35 \times 10^4)$ in dichloroethane. Specific Kerr constants (B/c) for the two samples were obtained from the plots of the Kerr constant versus concentration. A molecular length for the high molecular weight sample was obtained from the rotational diffusion constant at infinite dilution, using either a rod or ellipsoidal model. The shape of the rise and decay curves and the molecular weight dependence of the specific Kerr constant showed that both permanent and induced dipole orientation were important.

In 1959 O'Konski et al. [5] proposed a method for determining electric and optical parameters from saturation of the electric birefringence (see Chap. 3), and applied it to PBLG. The steady state electric birefringence of PBLG in dichloroethane was measured as a function of field strength, up to 3×10^4 V/cm. The data could be fitted to the theoretical equation for the case of pure permanent dipole orientation of axially symmetric macromolecules. The permanent dipole moment per amino acid residue was found to be about 3 D. The optical anisotropy factor $(g_1 - g_2)$, computed from the birefringence extrapolated to infinitely high field strength, was 4.0×10^{-3}, in agreement with the value obtained from streaming birefringence.

In the same year, Wada [6,7] investigated the dielectric dispersion of a homologous series of PBLG, ranging in molecular weight from 7×10^4 to 18×10^4, in dichloroethane. The permanent dipole moment, calculated from the specific polarization extrapolated to zero frequency, was proportional to the degree of polymerization and the permanent dipole moment per amino acid residue was found to be 3.4 D. The relation between the critical frequency and the degree of polymerization was elucidated, using an equivalent ellipsoid model and the Perrin equation for the rotational diffusion constant. These facts confirmed that PBLG had the rigid α helical structure with an ordered arrangement of residues, at least in this solvent and in the molecular weight range studied.

Yamaoka [8] carried out an extensive study on the field strength dependence of the electric birefringence of PBLG in various solvents, including tetrahydrofuran, dichloroethane, chloroform, dimethylformamide, dioxane, and dichloroethane-dichloroacetic acid mixtures. The permanent dipole moment, the anisotropy of electric polarizability, and the optical anisotropy factor were separately determined by analyzing the saturation curves. Also, molecular lengths were obtained from the decay curves. Thus, the permanent dipole moment and the axial translational length per amino acid residue were calculated for the PBLG molecule in each solvent. These data give a clue to the type of helical conformation in solution, such as α helix or 3_{10} helix. Furthermore, the wavelength dependence of the optical anisotropy factor was studied.

Boeckel et al. [9] measured the electric birefringence of three PBLG samples, ranging in molecular weight from 4×10^4 to 3×10^5, in pulsed and sinusoidal fields. The solvents used were

dichloroethane and a dichloroethane-dimethylformamide mixture
(99:1). Dimethylformamide was added in order to prevent aggregation of PBLG molecules. The birefringence tended to zero as the
frequency of the sinusoidal field was increased, as expected for
the mechanism of permanent dipole orientation. Both the permanent
dipole moment and the axial translational length per amino acid
residue decreased with an increase of molecular weight. This can
be explained only in terms of some chain flexibility.

Tsvetkov et al. [10-12] studied the electric, optical, and
hydrodynamic properties of a series of PBLG fractions in
dichloroethane containing 1% dimethylformamide, and in m-cresol by
combining the electric birefringence in a sinusoidal field (for a
range of frequencies from 20 Hz to 150 kHz) with the streaming
birefringence. The molecular weight range covered was from
6×10^4 to 32×10^4. The experimentally found dependence of the
permanent dipole moment and the rotational diffusion constant upon
the molecular weight indicates that the conformation of the PBLG
molecules is intermediate between a rectilinear rod and a Gaussian
coil.

Now, the mean square dipole moment of the wormlike chain of
Kratky and Porod [13] is given by

$$\overline{\mu^2} = 2\mu_0^2\left[\frac{L}{a} - 1 + \exp\left(\frac{-L}{a}\right)\right] \tag{1}$$

where L is the contour length, a is the persistence length,
and μ_0 is the dipole moment of the straight portion of the chain
having a length a. Comparison of the molecular weight dependence
of the dipole moment with this theoretical relationship yielded a
persistence length of 700 Å for PBLG. On the other hand, the
value of persistence length obtained from the molecular weight
dependence of the streaming birefringence was 1200 Å [14,12].
This disagreement means that the processes of orientation of PBLG
molecules in electric and hydrodynamic fields are not identical.

Watanabe et al. [15] measured the specific Kerr constant and the dielectric constant of PBLG in dichloroethane-dichloroacetic acid mixtures as functions of the solvent composition. These properties underwent a marked decrease with the addition of a small amount (less than 1%) of dichloroacetic acid, as well as an abrupt change accompanying the helix-coil transition in the vicinity of 75 vol% dichloroacetic acid. The former was attributed to the apparent diminution of the dipole moment of PBLG molecules, which might be due to the protonation of terminal amide groups. Furthermore, the specific Kerr constant of PBLG in 75 vol% dichloroacetic acid was measured as a function of temperature. The specific Kerr constant rapidly increased in a relatively narrow temperature range. This behavior was similar to the temperature dependence of the optical rotation, and was attributed to the transition of coil into helix.

Subsequently, Watanabe and Yoshioka [16] studied the effect of various acids on the specific Kerr constant of PBLG in dichloroethane. The addition of a small amount of strong acids (trifluoroacetic acid, trichloroacetic acid, monochloroacetic acid, and hydrogen chloride) caused a rapid decrease of the specific Kerr constant. On the other hand, the effect of weak acids (formic acid, acetic acid, and propionic acid) was small. The electric conductivity of PBLG in dichloroethane-dichloroacetic acid mixtures increased considerably at small fractions of dichloroacetic acid. These facts support the hypothesis given above. They also studied the effect of dichloroacetic acid on the specific Kerr constant of PBLG in other helicogenic solvents, such as m-cresol and dimethylformamide.

Tsvetkov et al. [17] measured the characteristic optical anisotropy $\Delta n/G(\eta - \eta_0)$ (where Δn is the streaming birefringence, G is the velocity gradient, η and η_0 are the viscosities of the solution and the solvent, respectively), the specific Kerr constant, and the reduced viscosity of PBLG in dichloroethane-dichloroacetic acid mixtures as functions of the solvent

composition. There were two regions of the solvent composition
where the characteristic optical anisotropy and the specific Kerr
constant suffered distinct changes. The decrease of these quantities in the region of 65-75% dichloroacetic acid corresponds to
the helix-coil transition. The other sharp decrease in the
region of 0.1-10% dichloroacetic acid was related to the drop of
molecular optical anisotropy and effective dipole moment due to
local conformation changes in helical chains of PBLG, which was
not revealed on the intrinsic viscosity.

Also, Tsvetkov et al. [18] measured the streaming birefringence and the electric birefringence of PBLG in dichloroacetic
acid. The values of the optical anisotropy, the rotational diffusion constant, and the dipole moment of PBLG in helix and coil
conformations were compared. It was concluded that the rigidity
of helical molecules was by some 100 times higher than that of
random coil.

Nishinari and Yoshioka [19] proposed a theory for the rise of
the electric birefringence of rigid macromolecules in solution,
which holds for arbitrary high field strength in the initial
stage. According to this theory, it is possible to determine the
permanent dipole moment and the anisotropy of electric polarizability separately from the field strength dependence of the initial
slope of the rise curve ($\Delta n(t)$ vs. t) and the initial slope of
the $\Delta n(t)/t$ vs. t curve. This method was applied to PBLG in a
helical conformation [20]. The electric birefringence of PBLG in
dichloroethane containing 1% dimethylformamide under the action of
a rectangular pulse was measured over a wide range of field
strength of 1.5×10^{3} to 3×10^{4} V/cm. The $\Delta n(t)/t$ vs. t
curves for various field strengths were linear in the initial
stage and passed through the origin. Thus, it was concluded that
the electric orientation of PBLG was due primarily to the permanent dipole moment.

The stability of PBLG helix at high temperatures was studied
by Tsuji et al. [21]. They measured the electric birefringence

and viscosity of PBLG in m-cresol up to 170 and 200^{0}C,
respectively. No abrupt changes which suggest breakage of the
helical structure were observed. However, the PBLG helix became
considerably more flexible with increasing temperature.

The effect of dichloroacetic acid on the electric and
hydrodynamic properties of PBLG in dichloroethane has recently
been reinvestigated by Ohe et al. [22]. In this study the
apparent dipole moment and the optical anisotropy factor were obtained from the field strength dependence of the steady state
birefringence, and the average relaxation time was obtained from
the decay curve of birefringence. The experimental results suggest that the first addition of dichloroacetic acid gives rise to
dissociation of aggregates, protonation of terminal amide groups,
and changes in the average orientation of side chains. The most
important contribution to the initial abrupt drop of the specific
Kerr constant comes from the second factor.

Marchal et al. [23] studied the conformation of two kinds of
poly-γ-benzyl-DL-glutamate, one polymerized in benzene and the
other polymerized in dioxane or dimethylformamide, by measuring
the dielectric dispersion and the electric birefringence. The
former behaved like a statistical coil of rodlike helical units
and the behavior of the latter was quite similar to that observed
for PBLG.

C. Electric Birefringence of Polyglutamic Acid

Yamaoka [8] measured the field strength dependence of the
electric birefringence of poly-L-glutamic acid (PLGA) in
dimethylformamide. The results indicate that PLGA exists in a
helical conformation with a large permanent dipole moment along
the helix axis in this solvent.

Matsumoto et al. [24] studied the electric birefringence of
PLGA in methanol-water mixtures. The permanent dipole moment, the

optical anisotropy factor, and the molecular length of PLGA were
obtained from the field strength dependence of the steady state
birefringence and the analysis of decay curves. The increase of
water content, from 0 to 40 vol%, caused a decrease of the permanent dipole moment and the molecular length. This suggests that
the helical PLGA molecules become partly flexible upon the addition of water. The aggregation of PLGA molecules was observed in
70, 80, and 90 vol% water.

They developed a numerical method for determining the distribution function of relaxation times from the decay curve of
electric birefringence in polydisperse systems [25]. The distribution function is expanded in a series of appropriate functions
and the best values of the coefficients are determined by the
method of least squares. The distribution of molecular lengths
can be obtained from the relaxation spectrum in the case of helical polypeptides. Moreover, they proposed a method of obtaining
the mean permanent dipole moment and the optical anisotropy factor
from the field strength dependence of the birefringence in polydisperse systems on the basis of the knowledge on the length
distribution [26].

The conformation of PLGA in various organic solvents (i.e.,
methanol, dimethylformamide, dimethylsulfoxide, and dioxane-water
mixtures) was studied by applying these methods [26]. The results
show that PLGA in methanol and dimethylformamide is presumably in
the α-helical conformation and PLGA in dimethylsulfoxide may
exist in a helical conformation with the length per amino acid
residue shorter than that of α-helix. Indications of solvent-
induced and temperature-induced transitions between the two helical forms were observed with PLGA in dimethylsulfoxide-methanol
mixtures [27]. The specific Kerr constant of PLGA in
trifluoroacetic acid was very small, which suggests that PLGA is a
random coil in this solvent.

Powers [28,29] measured the Kerr constant of the acridine
orange-PLGA complex in dimethylformamide as a function of

wavelength through an entire region of the visible absorption.
The spectrum of the complex in dimethylformamide showed a maximum
at 4950 Å and a shoulder at 4650-4700 Å. The Kerr constant
exhibited anomalous dispersion, and the form of this dispersion
was related to the direction of the transition moment for the
particular electronic transition. It was concluded that the long
axis of the bound dye was approximately parallel to the long axis
of the PLGA helix. The information was complementary to that
available from measuring the circular dichroism of partially
oriented solutions of the complex.

D.   Electric Birefringence of Polylysine

Poly-L-lysine (PLL) is a polypeptide with ionizable side
chains. PLL hydrochloride or hydrobromide undergoes a helix-coil
transition between 87 and 90 vol% methanol at neutral pH in
methanol-water mixtures [30]. In this case the ε-amino groups of
PLL are charged even in the helical conformation, as judged from a
nuclear magnetic resonance (NMR) study [31]. Thus, PLL serves not
only as a model for flexible polyelectrolytes, but also for rodlike polyelectrolytes.

Kikuchi and Yoshioka [32] studied the electric birefringence
of poly-L-lysine hydrobromide in methanol-water mixtures ranging
from 0 to 98 vol% methanol. An abrupt change in the specific Kerr
constant accompanied the solvent-induced helix-coil transition.
The specific Kerr constant increased rapidly with dilution in the
random coil form, and more slowly in the helical conformation.
The field strength dependence of the birefringence, for both the
helical and coil conformations, could be described by a common
orientation function, which resembled the theoretical one for
permanent dipole moment orientation. This was also the case for
several fractions of potassium polystyrene sulfonate in aqueous
solution [33]. These results were interpreted in terms of the
saturation of ion-atmosphere polarization proposed by Neumann and

Katchalsky [34].

It is worthy of special mention that anomalous birefringence signals were observed above a threshold field strength in the vicinity of 90 vol% methanol [32]. The threshold field strength varied with the polymer concentration and the solvent composition; it was about 5 kV/cm for PLL hydrobromide $(M = 1.4 \times 10^5)$ at a concentration of 1.4×10^{-4} g/cm^3 in 90 vol% methanol. A typical oscilloscope trace of the anomalous signal is shown in Fig. 1. Upon application of a rectangular pulse, the birefringence passed through a maximum and began to decrease slowly before the pulse terminated, reaching a steady-state value. The anomalous signals were well reproduced when the solution was pulsed repeatedly. This was interpreted as indicating that high electric fields provoke a transition from the charged helix to the charged coil in

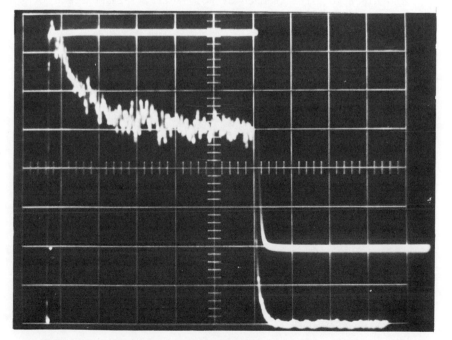

FIG. 1. Anomalous birefringence signal for poly-L-lysine hydrobromide in 90 vol% methanol. Concentration: 1.4×10^{-4} g/cm^3. Field strength of applied pulse: 10.1 kV/cm. Horizontal time scale: 0.1 msec per division.

the transition region. Another evidence in support of the
field-induced helix-coil transition is that the steady state
birefringence vs. E^2 curve for PLL hydrobromide in 90 vol%
methanol levels off above a threshold field strength, tending to
give a saturation value closer to that of the coil form.   This
experiment shows that electro-optical methods are promising for
the study of electric field induced conformational changes of
biopolymers, as described in Chap. 3.

        E.   Electric Birefringence of Other Polypeptides

     Some fragmentary studies of the electric birefringence of
polypeptides other than PBLG, PLGA, and PLL have been reported.
     The electric birefringence of poly-β-benzyl-L-aspartate in
chloroform was considerably smaller than that of PBLG in the same
solvent [8].
     The temperature-induced helix-coil transition of poly-γ-
methyl-L-glutamate in a dichloroethane-dichloroacetic acid mixture
(35:65 by volume) and in a m-cresol-dichloroacetic acid mixture
(35:65 by volume) was followed by measuring the electric birefringence [35].
     Poly-L-alanine exhibited a large specific Kerr constant in
dichloroacetic acid, corresponding to a helical conformation [36].
The specific Kerr constant decreased with the addition of
trifluoroacetic acid.
     Tsvetkov et al. [18] measured the electric birefringence and
the streaming birefringence of poly-ε-carbobenzoxyl-L-lysine in
dimethylformamide and dichloroacetic acid.
     An electric birefringence study of poly-L-proline will be
described in the following section.

F. Aggregation of Polypeptides
Studied by Electric Birefringence

Polypeptide molecules tend to aggregate more or less easily
in solution, and the aggregation is often accompanied by large
changes of electric and hydrodynamic properties. Therefore,
electric birefringence serves as a powerful method for studying
the nature of the aggregates of polypeptides.

Watanabe [37] studied the molecular association of PBLG in
benzene, dioxane, chloroform, and dichloroethane, and also in mixtures of these solvents with dimethylformamide or dichloroacetic
acid by measurements of the specific Kerr constant and the birefringence relaxation time. It was concluded that the association
of PBLG in benzene or dioxane was of a side-by-side type with
antiparallel orientation and the association in chloroform or
dichloroethane was of a linear type with head-to-tail orientation.
The aggregates in these solvents were dissociated by the addition
of a small amount (1-5 vol%) of dimethylformamide or dichloroacetic acid. However, in benzene or dioxane, they were not completely dissociated until the fraction of dimethylformamide or
dichloroacetic acid reached 40 to 60%. When the fraction of
ethanol was increased in dichloroethane-ethanol and dimethylforma-
mide-ethanol mixtures, PBLG was precipitated. The association of
PBLG in dichloroethane-ethanol mixtures was a linear type with
head-to-tail orientation and the association in dimethylformamide-
ethanol mixtures was a side-by-side type with antiparallel orientation.

Powers and Peticolas [38] investigated the mode of aggregation of PBLG dissolved in benzene, dioxane, and dichloroethane by
measuring the Kerr constant as a function of concentration. From
a consideration of this concentration dependence, the calculated

dipole moments, and the measured intrinsic viscosities, two
different modes of aggregation were proposed -- a linear association for high dielectric solvents (dichloroethane) and a lateral
(antiparallel) association for low dielectric solvents (benzene,
dioxane).

Subsequently, they studied the aggregation of PBLG by measuring the specific Kerr constant over a tenfold range of concentration and the intrinsic viscosity of solutions of the polymer in
four solvent mixtures, i.e., benzene-dimethylformamide, benzene-
dichloroethane, dioxane-dimethylformamide, and dioxane-dichloro-
ethane [39]. A low molecular weight sample $(M = 3.7 \times 10^4)$ was
selected in order to maximize any effects due to the number of
polymer ends in solution. Sharp changes were found in these
quantities on the addition of small amounts of polar solvent to
solutions of the PBLG in either benzene or dioxane; this implies
that lyotropic phase changes are occurring. There was evidence
for a birefringent and highly viscous phase in benzene solution
(probably smectic in nature), but in dioxane the aggregation
appeared to be only short-range and predominantly antiparallel.
In the benzene-dimethylformamide system the birefringent phase
changed to one similar to that found in dioxane before complete
dissociation occurred.

Matsumoto et al. [40] studied the aggregation of PLGA in
aqueous methanol solvents. The rise and decay of the electric
birefringence for PLGA in aqueous solvents containing 20 and 10
vol% methanol was found to be unusual. For example, the birefringence changed sign in the decay process under some conditions.
The decay curves were analyzed on the assumption that there exist
two kinds of particles, namely, one (component I) with a shorter
relaxation time exhibiting positive birefringence and the other
(component II) with a longer relaxation time exhibiting negative
birefringence at low fields. From the field strength dependence

of the steady state birefringence the permanent dipole moment, the
anisotropy of electric polarizability, and the saturation value of
electric birefringence were determined for each component.
Furthermore, from the relaxation time the length of component I
and the diameter of component II were calculated on the models of
cylindrical rod and oblate ellipsoid, respectively. The permanent
dipole moment, the anisotropy of electric polarizability, and the
relaxation time of component II were much larger than those of
component I. Both the anisotropy of electric polarizability and
the optical anisotropy factor were positive in sign for component
I and negative for component II. It was concluded that component
I was the helical PLGA molecule itself and component II was the
side-by-side (antiparallel) aggregate composed of many helical
PLGA molecules, of the order of 10^4.

   Brown and Jennings [41] studied the aggregation of poly-L-
proline II (trans configuration) in aqueous solution by means of
electric birefringence. Poly-L-proline II has the tendency to
precipitate at room temperature. This precipitation is apparently
enhanced by increase of temperature and checked or reversed by
lowering the temperature to 5°C. To investigate the temperature
dependence of the aggregation, an undialyzed solution of
1×10^{-3} g/ml was cooled to 5°C. Transient electric birefringence traces were recorded at various temperatures. The relaxation time was shorter, the lower the temperature. The results
showed the disruption of the aggregate upon cooling. By comparing
the area above the rise curve with the area below the decay curve
according to the method of Yoshioka and Watanabe [42], it was
demonstrated that the permanent and induced dipole moments of the
aggregate formed at 25°C would be in quadrature. A large raftlike
or disc-shaped aggregate of laterally packed polyproline helices
was suggested.

G.   Electric Dichroism of Polypeptides

Electric dichroism of some polypeptides in helical
conformations has been measured in the infrared and ultraviolet
regions.   This method can provide information about the direction
of chromophoric transition moments with respect to the axis of
orientation (e.g., the helix axis).   In addition, the permanent
dipole moment and the rotational diffusion constant are determinable.

In 1959 Spach [43] studied the infrared electric dichroism of
PBLG in chloroform or dioxane.   Infrared absorption spectra of
PBLG in solution were compared in the absence and presence of an
electric field.   The electric field was applied in the direction
perpendicular to the incident radiation, and the radiation was
linearly polarized so that its electric vector was parallel to the
field direction.   The intensity of the amide A band (at 3300 cm^{-1},
N-H stretching) and the amide I band (at 1655 cm^{-1}, C=0 stretching) increased; on the other hand, that of the amide II band (at
1550 cm^{-1}, N-H in-plane bending and C-H stretching) decreased.
The results offer further evidence that PBLG preserves the α-
helical conformation in appropriate solvents.

About the same time, Miyazawa and Blout [44] studied the
infrared dichroism of PBLG in dichloroethane by applying an electric field in the direction parallel to the incident, unpolarized
radiation.   The cell was made of a pair of calcium fluoride
windows with Nesa-coated inner surfaces.   The intensity of the
amide A band decreased in the presence of the electric field.   The
fractional decrease of the intensity of this band was related to
the degree of orientation of the helical PBLG molecules in solution.

Quantitative measurements of the infrared electric dichroism
of PBLG solutions have been performed by Maschka et al. [45] and
by Champion and Nicholas [46].   The former authors determined the
direction of transition moments from the dichroic ratio at the

amide A band and the amide I band of PBLG in chloroform, using a
relation between the field strength and the degree of orientation.
The latter authors measured the dichroism of the amide II band (at
1550 cm^{-1}) of PBLG (M $\approx$ 3.0 $\times$ 10^5) in dioxane as a function of
field strength, up to 4 $\times$ 10^4 V/cm. In order to obtain correspondence between theory and experiment, polydispersity of the
sample and flexibility of the helix had to be taken into account.

Next, we shall turn to the ultraviolet dichroism. Shirai
[47] measured the ultraviolet electric dichroism of PBLG in
tetrahydrofuran at 263 nm, where the absorption due to benzyl
groups predominates. A sign change of the dichroism from positive
to negative was observed as the temperature was changed from 18 to
5°C.

Charney et al. [48] studied the ultraviolet electric dichroism
of PBLG in solution. The wavelength covered was from 245 to 285
nm. The peak of the absorption band is located at 258 nm. This
absorption is due to the benzyl group of the side chain. The re-
duced dichroism, defined as $\Delta\varepsilon/\varepsilon = (\varepsilon_\parallel - \varepsilon_\perp)/\varepsilon$, where $\varepsilon_\parallel$ and
$\varepsilon_\perp$ are the molar extinction coefficients parallel and perpendicu-
lar to the electric field, respectively, and ε is the isotropic
molar extinction coefficient, was measured as a function of field
strength and wavelength in several solvents (i.e., dichloroethane,
dichloromethane, and dioxane). The dichroism increased with increasing field strength, and saturation was reached in dichloroethane. In dichloroethane at 258 nm, the value of $\Delta\varepsilon/\varepsilon$
extrapolated to infinitely high field strength was +0.09. For a
rigid dipolar macromolecules, the limiting value of $\Delta\varepsilon/\varepsilon$ is related to the angle between the transition moment of the chromo-
phore and the permanent dipole moment, ξ, by the equation

$$\left(\frac{\Delta\varepsilon}{\varepsilon}\right)_{E\to\infty} = \frac{3}{2} (3 \cos^2 \xi - 1) \tag{2}$$

Thus, it was estimated that the electronic transition moment for
the 258 nm absorption band made an angle of 53.5° with the helix

axis (i.e., the direction of the permanent dipole moment). With
this information a configuration of the side chain of PBLG was
proposed which was in accord with infrared data from solid films.
The dependence of the electric dichroism on the field strength in
dichloroethane and dichloromethane was largely linear even at low
fields. This indicates that even at molecular weights of 100,000,
the lowest molecular weight polymer measured, the PBLG molecule is
flexible, although the flexibility does not destroy the rodlike
characteristics.

No measurable electric dichroism at 258 nm was observed with
poly-β-benzyl-L-aspartate (with one less methylene group on the
side chain). This indicates that either the side chains are
randomly oriented or are so oriented as to reduce the overall dipole moment of the polymer to too small a value to be measured.

Milstien and Charney [49] studied the effect of trifluoroacetic acid on the electric dichroism of PBLG in dichloroethane at
258 nm. The addition of small amounts (up to 5%) of trifluoroacetic acid caused complete disappearance of dichroism in contrast to
the electric birefringence, which dropped to an observable plateau
at 20% of its initial value. This shows that the initial effect
of strong acid is on the side chain, not on the helix backbone.
In order further to examine the nature of interaction of
trifluoroacetic acid with the side chain, ultraviolet absorption,
infrared absorption, and NMR measurements were carried out. The
results indicate that trifluoroacetic acid in small amounts interact with the ester carbonyl oxygen in the side chain, either by
hydrogen bonding or protonation.

Troxell and Scheraga [50,51] developed a new, highly sensitive method for measuring the electric dichroism of macromolecular
solutions. In this method, the rotation of the plane of incident
linearly polarized light by a solution is measured by a
spectropolarimeter in the absence and presence of an electric
field. The applied static electric field is perpendicular to the
direction of the incident light. The angle between the major axis

of the output elliptically polarized light and the field direction
is set at $45°$. Then, the change of the rotation in degrees produced by the electric field is given by the approximate formula

$$\alpha_{E,45°} \equiv (\alpha_{E\neq0} - \alpha_{E=0})_{45°} = 33(\varepsilon_\parallel - \varepsilon_\perp)lC \qquad (3)$$

where l is the pathlength in medium and C is the molar concentration. When the electric orientation of the macromolecules is
due solely to the permanent dipole moment μ, the values of
$(3 \cos^2 \xi - 1)/2$ and μ are obtained from the initial slope and
intercept of an $\alpha_{E,45°}/E^2$ vs. E^2 plot. Representative data
for poly(p-chloro-β-benzyl-L-aspartate) and poly(p-chloro-γ-
benzyl-L-glutamate) in dioxane confirmed the validity of this
method.

As a preliminary study, the electric dichroism spectra of
poly-L-tyrosine (degree of polymerization, $\sim$200) in dioxane were
measured in the wavelength range of 210 to 300 nm according to
this technique [50]. A value of μ of 1090 D was obtained; this
is consistent with the conclusion drawn from circular dichroism
and optical rotatory dispersion measurements that the polymer is
highly helical. The calculated values of $(3 \cos^2 \xi - 1)/2$ were
small, being +0.042 at 279.6 nm and -0.043 at 227 nm. These
values were interpreted in terms of the orientation of the tyrosyl
group of the side chain. This electric dichroism technique was
also applied to poly(n-butyl isocyanate), which has the repeating
unit $-N(CH_2CH_2CH_2CH_3)-CO-$ and assumes a structure similar to
that of α-helical polypeptides [52].

Jennings and Baily [53,54] proposed a "longitudinal" method
of measuring the electric dichroism in pulsed fields. A conventional spectrophotometer was used apart from the special cell de-
sign. The cell had optically transparent end windows, coated with
a conducting material, which acted as the electrodes. The electric field was applied to a solution in the cell along the incident light, which need not be polarized at all. The resulting
absorption change should be identical to that in the case where

the incident light is polarized perpendicular to the applied field. Transient traces were obtained from a variety of polymers, including PBLG, in solution in the visible and ultraviolet regions.

Iizuka [55,56] studied the orientation of liquid crystals of PBLG in electric fields by measuring the infrared electric dichroism at 3300 cm^{-1} (due to N-H stretching vibration). In this study the dichroism was expressed by the dichroic ratio, $A_{\parallel}/A_{\perp}$. PBLG was dissolved in dibromomethane or dichloromethane at a concentration over 6%, and was allowed to stand at room temperature for at least 1 week. The solution was then put into a quartz cell of pathlength 1 mm, with a quartz spacer to adjust the pathlength to 0.10 mm and a pair of needle electrodes. When a liquid-crystalline structure was present in the solution, the dichroism was produced by an electric field as low as 84 V/cm. Such a good orientation was attributed to some cooperative behavior of the polypeptide molecules. The group of molecules that behaved cooperatively was tentatively named a molecular cluster. The dichroic ratio of the liquid-crystalline solutions increased with time after application of a static electric field until the steady state was attained. Effects of solute concentration, temperature, and field strength upon the feature of the time dependence were investigated. The steady state dichroic ratio increased with increasing field strength, the asymptotic value being 4.5-4.7 regardless of the solute concentration when it ranged from 13.5 to 25.3% (v/v) where only the birefringent phase existed in solution. In dilute liquid-crystalline solutions, the PBLG molecules behaved as if independent molecular clusters having a dipole moment some 730 times as large as that of the single PBLG molecule (of degree of polymerization 650) were present in solution. In addition, the electric birefringence of the liquid-crystalline solutions of PBLG was measured. The features of the time dependence and the field strength dependence were analogous to those observed for the dichroic ratio.

The electrical orientation of liquid crystalline solutions of poly-γ-ethyl-L-glutamate has also be studied by infrared dichroic ratio [57].

H. Electric Field Light Scattering of Polypeptides

When application of an electric field leads to a partial orientation or deformation of the macromolecules, comparable in size to the wavelength of visible light, a change in the intensity of the scattered light is produced at all angles of observation. Conventional light scattering method is widely used for determining the molecular weight and the radius of gyration of macromolecules in solution. Light scattering subjected to electric fields yields additional information about the electric properties (permanent dipole moment and anisotropy of electric polarizability), the rotational diffusion constant, and the molecular flexibility (see Chap. 9).

As early as in 1956, Wippler [58,59], founder of the electric field light scattering method, studied the effect of electric field on the light scattering of poly-DL-phenylalanine in benzene. The intensity of the scattered light was measured at 90° to the direction of the incident light. The directions of the applied field were such that $\Omega = 0$ and 90°, where Ω is the angle between the field direction and the vector defined by $s = e_0 - e$, e_0 and e being unit vectors in the direction of the incident and scattered light, respectively. The intensity change due to the field was almost zero for $\Omega = 90^{\circ}$, and negative for $\Omega = 0$. This indicates that poly-DL-phenylalanine molecules behave as statistical coils in benzene. The permanent dipole moment of the chain element per unit length was calculated to be 0.49 D from the field strength dependence of the intensity change.

In 1962 Wallach and Benoit [60] studied the light scattering of three samples of PBLG, ranging in molecular weight from

1.8×10^5 to 3.2×10^5, in dichloroethane when subjected to electric fields. The intensity change in the dc field was measured as a function of the field strength, the angle of observation, and the direction of the field. The ratio of the intensity change produced by the field applied in two specified directions, $\Omega = 0$ and 90^0, was close to -2. This shows that the helical PBLG molecules are not deformed by the field. Dispersion measurements in the ac field indicate that the electric orientation is due primarily to the permanent dipole moment. The rotational diffusion constant, obtained from the critical frequency, was found to vary inversely as the cube of the rod length, as predicted by the theory of Burgers or Broersma. The permanent dipole moment per amino acid residue was calculated to be 3.5 D, in agreement with the data obtained from electric birefringence and dielectric measurements.

Subsequently, Jennings and Jerrard [61] made a similar study on the electric field light scattering of PBLG in dichloroethane. The weight-average molecular weight of the PBLG sample was 2.5×10^5. The ratio of the z-average length of the equivalent rod to the weight-average degree of polymerization was found to be 1.36 Å. The solutions were subjected to dc fields up to 1.9 kV/cm and ac field of 450 V/cm at frequencies up to 10 kHz. The intensity change decreased with increasing frequency until the asymptotic value was reached. The high-frequency asymptote, arising from the anisotropy of electric polarizability, was much smaller than the frequency dependent contribution due to the permanent dipole moment. The molecule was found to possess a permanent dipole moment of 3920 D, corresponding to 3.44 D per amino acid residue. The variation of the rotational diffusion constant with molecular weight, found from the present results and those obtained by Wallach and Benoit, indicate the possibility that the PBLG molecule could be rodlike and flexible.

I. Electric-streaming Birefringence of Polypeptides

In 1962 Mukohata et al. [62,63] developed a new method,
which might be called the electric-streaming birefringence method,
for estimating the electric and optical properties together with
the rotational diffusion constant of macromolecules by means of a
single experimental technique.

The theoretical basis of this method was established by Ikeda
[64]. The angular distribution function of rigid macromolecules
in solution was obtained by solving the diffusion equation in the
steady state, when the solution flows with a constant velocity
gradient and is exerted by a static electric field perpendicular
to the stream line. The macromolecule was assumed to be an ellipsoid of revolution with a permanent dipole moment along the symmetry axis. By the use of this distribution function, the
extinction angle and the magnitude of the birefringence were expressed as functions of the velocity gradient and the field
strength. When both the velocity gradient G and the field
strength E are small, the extinction angle χ is given by

$$\cot 2\chi = \frac{G}{6\Theta} - \frac{\Theta}{RG}\left(\frac{\mu^2}{k^2 T^2} + \frac{\alpha_1 - \alpha_2}{kT}\right)E^2 + \ldots \tag{4}$$

where Θ is the rotational diffusion constant around the transverse axis and R is a geometrical parameter defined by
$R = (p^2 - 1)/(p^2 + 1)$, p being the axial ratio.

The electric-streaming birefringence method was first applied
to PBLG $(M = 20.6 \times 10^4)$ dissolved in m-cresol [62]. A streaming birefringence apparatus of the concentric cylinder type was
used, and a dc field was applied between the two cylinders. The
extinction angle was measured for various velocity gradients and
field strengths. The rotational diffusion constant was obtained
from the initial slope of the χ vs. G curve in the absence of

the electric field. A plot of cot 2χ against E^2 was linear if
G was constant, as predicted by the theory. From the slope of
the line and the value of Θ, the electric factor,
$\mu^2/k^2T^2 + (\alpha_1 - \alpha_2)/kT$, was calculated, putting R equal to
unity. In the case of helical PBLG molecule, the anisotropy of
electric polarizability, $\alpha_1 - \alpha_2$, may be neglected. Thus, the
permanent dipole moment per amino acid residue was found to be
3.4 D.

Further experiments were carried out on the same system by
Mukohata [65], using an ac field instead of a dc field. The
extinction angle measured at a given G and E changed with the
frequency of the applied field. The rotational diffusion constant
was obtained from the critical frequency.

Tsvetkov and Vinogradov [66] studied the birefringence and
the extinction angle of solutions of PBLG fractions, ranging in
molecular weight from 6×10^4 to 33×10^4, under the joint action
of a laminar flow and a perpendicular electric field. The solvent
used was a chloroform-dimethylformamide mixture (97.7:2.3 by volume). The frequency of the applied field was 50 Hz and the maximum field strength was 3×10^3 V/cm. This method yielded
quantitative data on hydrodynamic, electric, and optical properties of PBLG. The results were interpreted on the basis of the
theory of birefringence of wormlike chains.

II.   ELECTRO-OPTICS OF PROTEINS

A.   Introduction

It is customary to distinguish, on the basis of the molecular
shape, two classes of proteins: fibrillar and globular. Fibrillar
proteins, such as fibrinogen, collagen, and muscle proteins, in
solution exhibit large electric birefringence, as well as large
streaming birefringence. Their birefringence relaxation times are
of the order of 100 μsec. These proteins are suitable for

transient electric birefringence studies. On the other hand, the
specific Kerr constants of globular proteins, such as albumins and
globulins, are much smaller; their relaxation times lie in the
submicrosecond region. It was not until 1957 that the transient
electric birefringence of some globular proteins in solution was
successfully measured with the introduction of a high-voltage,
high-speed delay line pulse generator (see Sec. F).

The electric birefringence has been applied to the study of
the dimensions and electric properties of some proteins in solution, and also to the study of changes in these properties under
various circumstances, as will be described in the following sections.

The electric dichroism of protein solutions has not yet been
reported, as far as the author knows at the end of 1973. However,
it will become a valuable means for investigating the spatial
arrangement of chromophore groups within protein molecules in the
solution state, in parallel with the streaming dichroism currently
in use.

B.   Electric Birefringence of Fibrinogen

In 1954 Tinoco [67,68] studied the electric birefringence,
under the action of a rectangular pulse, of bovine fibrinogen in
3 M urea in 64% aqueous glycerol before and after activation by
thrombin. This was the first application of the transient electric birefringence method to proteins. The average rotational
diffusion constant, obtained from the decay curves and corrected
to the viscosity of water at $20^{\circ}C$, was 3.6×10^5 $\sec^{-1}$, in agreement with the results of streaming birefringence measurements.
The rise curves showed that at low pH (6 to 7) the orientation was
due mainly to the anisotropy of electric polarizability. However,
at higher pH (7 to 10) there were contributions from the permanent
dipole moment. The dipole moment, calculated from the specific

Kerr constant under the assumption that the optical anisotropy
factor was independent of pH in the same solvent, rose to a maximum of about 500 D at pH 8.5 and decreased again with increasing
pH. The dipole moment of activated fibrinogen was greater than
that of fibrinogen by an amount which varied with the pH. The
maximum change in dipole moment observed was 110 D. This difference, together with the fact that about 10 charges are lost on
activation, indicate that the site of attack by thrombin is near
the center of the fibrinogen molecule.

Subsequently, Billick and Ferry [69] studied the effect of
added urea on the electric birefringence of fibrinogen in salt-
free solutions (containing 64% glycerol) near pH 4.5 and 8.5.
Urea up to 3 M did not seriously affect the specific Kerr constant, and affected the rotational diffusion constant only through
modifying the protein interactions. At pH 8.5, the specific Kerr
constant (B/c) was about 8×10^{-3} cm^4 $statvolt^{-2}$ g^{-1}; at pH 4.5
it was about 5×10^{-3}. Activation by thrombin did not change the
rotational diffusion constant, but increased the specific Kerr
constant by about 1×10^{-3} cm^4 $statvolt^{-2}$ g^{-1} (at pH 4.5).

Haschemeyer and Tinoco [70] made a more critical study on the
transient electric birefringence of bovine fibrinogen in the pH
regions of 4.0 to 5.0 and 7.0 to 9.0. In part of the low pH
region, the occurrence of negative birefringence was observed. At
pH 4.5 the birefringence was negative at low fields; as the field
strength was increased, the birefringence reached a maximum negative value, then decreased and eventually changed sign. The negative birefringence was attributed to a species with a permanent
dipole moment along the transverse axis. The appearance of the
birefringence transients and anomalous titration behavior suggested the presence of more than one fibrinogen species in solution. The positive birefringence might be produced by other
species orienting parallel to the electric field by an induced

dipole mechanism. However, Holcomb and Tinoco [71] gave another
interpretation on the basis of the theory of birefringence
saturation extended by them to the most general molecular model,
assuming only one species present. In the high pH region the
presence of a longitudinal permanent dipole moment was confirmed.
The magnitude and the pH dependence of the specific Kerr constant
were consistent with the titration of two α-amino groups, one at
each end of the molecule, a distance of about 250 Å from the center. It was concluded that the distribution of groups titrating
in the pH regions studied was symmetrical about the center of the
fibrinogen molecule.

In the conversion of fibrinogen to fibrin monomer by thrombin
and by the snake venom extract Hemostase, peptide material is released and a protein intermediate is formed. Haschemeyer [72]
studied the structure of this intermediate and its kinetics of
formation by means of the transient electric birefringence. The
intermediate, which was identified as a fibrinogen molecule lack-
ing one A peptide, was characterized by a large longitudinal
permanent dipole moment. Its dipole moment was obtained as a
function of pH from birefringence measurements at saturating electric fields, and was used to determine the site at which charge
alteration had occurred. It was concluded that the two A peptides
of fibrinogen were located near the ends of the molecule, equidistant from the center. The presence of a transverse dipole moment
in fibrin monomer suggests that the sites may be on the same side
of the molecule. Comparison of rotational diffusion constants for
fibrinogen, the polar intermediate, and fibrin monomer showed that
little change in the length of the molecule occurred during peptide release. Direct determination of rotational diffusion constants during clotting at pH 8 established that the first step in
the polymerization of fibrin monomer was end-to-end dimerization.

C. Electric Birefringence of Collagen

Collagen is an elongated macromolecule consisting of three polypeptide chains wound in a rigid triple helix. Yoshioka and O'Konski [73] measured the transient electric birefringence of collagen dissolved in dilute acetic acid over a wide range of field strength. The soluble collagen used was from rat tail tendon. The specific Kerr constant (B/c) was very large, being 0.83 cm^4 statvolt^{-2} g^{-1} in 2.9×10^{-3} M acetic acid. Very pronounced saturation of birefringence was observed; at the highest field strength, 2×10^4 V/cm, the birefringence was close to complete saturation. The optical anisotropy factor, $g_1 - g_2$, was calculated from the birefringence extrapolated to infinitely high field strength. In 2.9×10^{-3} M acetic acid (pH around 4) the apparent permanent dipole moment and the anisotropy of electric polarizability, determined from the field strength dependence of the steady state birefringence, were 1.5×10^4 D and 2.7×10^{-15} cm^3, respectively. The contribution of the former to the specific Kerr constant was twice as large as that of the latter. The same conclusion was obtained also from the initial slope of the rise curves at low fields. The large longitudinal dipole moment of collagen may be caused by an asymmetrical distribution of charged groups in the molecule. The magnitude of the anisotropy of electric polarizability indicates that the ion atmosphere polarization is important. Moreover, effects of added salt and thermal denaturation on the electric birefringence were explored.

Kahn and Witnauer [74] measured the decay of the electric birefringence of calfskin collagen dissolved in citrate buffer at a number of concentrations. The log Δn vs. t plots were linear at concentrations ranging from 0.058 to 0.233%. This shows that the solubilized collagen preparation is monodisperse. At

higher concentrations the plotted lines became curved, which might
be due to either aggregation, interaction of collagen molecules,
or viscosity effects.

Subsequently, they studied the electric birefringence, including its rise and steady state, of calfskin corium collagen
dissolved in citrate buffer over the pH range of 3.25 to 4.85
[75]. The effects of collagen concentration, field strength, and
pH were investigated. At high concentrations the birefringence
transients (especially rise curves) showed anomalous behavior. In
some cases (e.g., collagen concentration = 1.05%, pH = 4.13,
E = 2,200 V/cm), the birefringence passed through a maximum, and
then decreased, even though the applied rectangular pulse was
still active. This was interpreted as indicative of time-dependent
variations in permanent and induced dipole moments. The permanent
dipole moment and the anisotropy of electric polarizability of
dissolved collagen were determined, using the same method as proposed by Yoshioka and O'Konski. The permanent dipole moment decreased and the anisotropy of electric polarizability increased as
pH was raised.

Ananthanarayanan and Veis [76] applied the electric birefringence method to characterization of pronase-treated acid-soluble
bovine skin collagen and untreated acid-soluble collagen in dilute
acetic acid. The former was characterized by a single decay process with rotational diffusion constant of 810 sec^{-1}. The decay
curves of the latter were interpreted to indicate the presence of
two species, monomer and end-linked dimer.

Recently, Bernengo et al. [77] have studied the electric
birefringence of acid-soluble calf skin collagen in 0.1 M acetic
acid, applying reversing pulses as well as rectangular pulses.
Reversing pulse experiments showed a very low contribution of induced moment compared to permanent dipole moment orientation.

D. Electric Birefringence of Muscle Proteins

Tropomyosin was discovered as the third structural protein in muscle in 1946. In 1961 Asai [78] measured the electric birefringence of rabbit tropomyosin under various conditions of solutions by the use of the rectangular pulse and the reversing pulse, for the purpose of investigating polymerization mechanism. On sudden reversal of the applied field, a large transient decrease of the birefringence was observed. This indicates directly the presence of a longitudinal permanent dipole moment. The specific Kerr constant and the average particle length, obtained from the birefringence decay by the use of Perrin's equation, decreased with dilution of the tropomyosin solution at various concentrations of added KCl or urea. At infinite dilution, the specific Kerr constant (B/c) was about 0.3 cm^4 $statvolt^{-2}$ g^{-1} and the length tended to 400 ± 50 Å at pH 6.9, regardless of the concentration of KCl or urea. This shows that the aggregated tropomyosin molecules are completely depolymerized to monomers by infinite dilution even under the salt-free condition. The permanent dipole moment of tropomyosin monomer was estimated to be about 390 D at neutral pH. Theoretical analysis made on the relation between the specific Kerr constant and the tropomyosin concentration leads to the conclusion that the polymerization is not only due to the head-to-tail association, but also due to the head-to-head or antiparallel side-by-side association. The binding constant between two tropomyosin molecules decreased with increasing concentration of KCl or urea. Above 30^0C the specific Kerr constant and the particle length decreased abruptly. This dissociation of tropomyosin molecules was completely reversible. The specific Kerr constant and the particle length were strongly dependent on pH at a fixed protein concentration. The specific Kerr constant attained a maximum at pH about 8, and remarkably decreased at pH lower than 6 and at pH higher than 9.

Actin monomers (G-actin) can be transformed into fibrous polymers (F-actin) by the addition of neutral salts. Kobayasi, Asai, and Oosawa [79] studied the polymerization of actin by measurements of electric birefringence. The polymerization was induced by a small amount of divalent cations (i.e., Ca^{2+}, Mg^{2+}). Small elementary polymers of G-actin were formed in the initial stage of polymerization of G-actin or by sonication of F-actin at low salt concentrations. These polymers showed a positive electric birefringence and a positive streaming birefringence. That is, they are oriented with the long axis parallel to the electric field. The large contribution of a longitudinal permanent dipole moment was confirmed by applying a reversing pulse. The magnitude of the dipole moment was estimated from the specific Kerr constant to be about 75 D per actin monomer. The elementary polymers grew into normal F-actin which showed a large negative electric birefringence and a large positive streaming birefringence. This means that F-actin filaments are oriented in the direction perpendicular to the electric field. The reversing pulse method suggested the presence of a transverse permanent dipole moment. The anomalous behavior of F-actin seems to contradict the helical polymer model. The negative electric birefringence of F-actin was cancelled by the addition of H-meromyosin. The electric structure of actin polymers may play an important role in the mechanism of muscular contraction.

In order to give stronger evidence for the transverse dipole moment of F-actin, Kobayasi [80] investigated the effect of pulsed fields on F-actin oriented by flow. A concentric cylinder was used as the Kerr cell, and F-actin in the cell was oriented by slow rotation of the inner cylinder. Then, a pulse was applied to the F-actin solution, and the change of birefringence was followed on an oscilloscope. The birefringence was decreased owing to the disorienting effect of the applied field. The birefringence decrease produced by an electric field parallel to the filament

axis of F-actin was much larger than that produced by a
perpendicular field. Moreover, a large transient increase of
birefringence was observed on reversal of a parallel field; on the
other hand, the reversal of a perpendicular field caused only a
small transient increase of birefringence. The analysis of the
birefringence change produced under the action of rectangular and
reversing pulses confirmed that F-actin had a large permanent dipole moment perpendicular to the filament axis. The possibility
of electrophoretic orientation was excluded. The marked disorienting effect of the parallel field on F-actin vanished on the addition of H-meromyosin. On dissociation of the complex between
F-actin and H-meromyosin by the addition of adenosine triphosphate,
the disorienting effect of the parallel field was again observed.
This result suggests that a strong electric interaction exists
between permanent dipoles of F-actin and H-meromyosin.

E. Electric Birefringence of Hemocyanin

Hemocyanin is a copper-containing blood pigment of mollusks
and other lower invertebrates. Pytkowicz and O'Konski [81]
measured the transient electric birefringence of Helix pomatia
hemocyanin in aqueous solution at $25^{0}C$. The solvents used were
10^{-4} M phosphate buffer (pH 7) and distilled water. The decay of
birefringence was not exponential in freshly prepared solutions
and indicated the presence of two relaxation times. The average
values were 11 and 53 μsec, corresponding to the rotational diffusion constants 15×10^{3} and 3.2×10^{3} sec^{-1}, respectively.
They interpreted the two relaxation times in terms of the whole
monomer and its end-to-end dimer.

Later, Ridgeway [82] showed that it was possible to give
another interpretation for these data. According to his theory
[83], the decay of birefringence for a monodisperse suspension of
asymmetric ellipsoids is describable by a sum of two exponential
terms. He assumed a single species of hemocyanin, instead of two,

and determined the axial lengths of an asymmetric ellipsoid of the
volume of the undissociated hemocyanin molecule which would lead
to the observed relaxation times. The dimensions of the anhydrous
molecule was calculated to be $38.4 \times 363.4 \times 1495$ Å on this model.

It is to be noted that Ridgeway says in his paper: "It is to
be emphasized, however, that there is no reason to reject the
original interpretation of Pytkowicz and O'Konski of their relaxation data as reflecting the presence of aggregated material found
in freshly prepared solutions which disappears upon long standing."
The asymmetric ellipsoid model is inconsistent with the cylindrical cross-section observed for linear aggregates of hemocyanin
[84].

F. Electric Birefringence of Some Globular Proteins
(Serum Albumin, Ovalbumin, β-Lactoglobulin,
γ-Globulin, Ribonuclease, Lysozyme, and Chymotrypsin)

In 1957 Krause and O'Konski [85,86] extended the transient
electric birefringence technique to the submicrosecond region for
structure studies of globular proteins and other smaller macromolecules, with lengths of the order of 100 Å, using a high-speed
delay line pulse generator which produced rectangular pulses of
1.7 μsec duration and variable amplitude to 8 kV. Bovine serum
albumin (BSA) was chosen as the first protein for investigation,
because it had been characterized extensively by other techniques.
The specific Kerr constant and the birefringence relaxation time
of BSA in aqueous solution were measured as functions of pH,
ionic strength, BSA concentration, and a variety of treatments.
The specific Kerr constants (B/c) of untreated and deionized BSA
both approached 2.4×10^{-4} cm^4 $statvolt^{-2}$ g^{-1} as the BSA concentration approached zero at $30°C$. The specific Kerr constant increased on either side of the isoelectric point (pH 5) and
decreased upon addition of electrolytes. The relaxation time at
1% of BSA in water at $30°C$ and pH 5 was 0.20 μsec. The

corresponding rotational diffusion constant was $8.3 \times 10^5 \ sec^{-1}$.
Combining this rotational diffusion constant with the known
molecular weight, partial specific volume and degree of hydration,
the dimensions of BSA were computed from Perrin's equations for
ellipsoids of revolution. A maximum dimension of 200 ± 35 Å was
obtained, assuming either an oblate or prolate ellipsoid; this
corresponds to an axial ratio far from unity in either case.
Comparison of these data with the previous results of dielectric
dispersion and fluorescence depolarization suggests that the prolate model is better than the oblate model. The birefringence
relaxation time decreased about 20% in going from pH 5.1 to 2.6.
This is consistent with an increase of molecular flexibility.

Furthermore, they measured the specific Kerr constant and the
birefringence relaxation time on γ-globulin in aqueous solution
and on BSA, β-lactoglobulin, ovalbumin, ribonuclease, lysozyme,
and chymotrypsin in glycerol-water solution [87]. The relaxation
times of γ-globulin and BSA could be observed in aqueous solution, but the relaxation times of the other proteins were observed
only in more viscous glycerol-water solutions. Rotational diffusion constants were calculated from the data and were compared
with those calculated from the protein dimensions obtained by
other methods, including dielectric dispersion, streaming birefringence, fluorescence depolarization, sedimentation and viscosity, and low angle x-ray scattering. The calculated and observed
rotational diffusion constants were in excellent agreement for
γ-globulin, ovalbumin, ribonuclease, and chymotrypsin, and were
within a factor of 2 for β-lactoglobulin. Lysozyme showed evidence of aggregation and BSA gave a lower relaxation time,
corrected for the solvent viscosity, in glycerol-water solution
than in aqueous solution, apparently because of a structural
change or a localized solvent viscosity effect.

Moser et al. [88] studied the dielectric dispersion and the
decay of electric birefringence on highly purified monomer of BSA
in water at various concentrations. Crystalline BSA was defatted,

deionized, and fractionated on a Sephadex column. The
birefringence relaxation time of the monomer, extrapolated to
infinite dilution, was 0.076 μsec at the isoelectric point at
25°C. Interpretation of the dielectric and birefringence relaxation times led to an axial ratio of 3.0 and a hydration of 0.64 g
per gram protein for a prolate ellipsoid model. All data were
consistent with a permanent dipole orientation mechanism. A
permanent dipole moment of 384 D was computed for the monomer.

In 1962 Ingram and Jerrard [89] constructed an electric
birefringence apparatus suitable for investigation of molecules
having relaxation times of the order of 5×10^{-8} sec and greater.
A detailed account of the apparatus was given in a subsequent
paper by Jerrard et al. [90]. Birefringence relaxation times for
a number of macromolecules were reported. Among them are included
BSA, bovine γ-globulin, and β-lactoglobulin. Then, two detailed
investigations on globular proteins followed.

Riddiford and Jennings [91] studied the conformation change
of BSA at low pH by transient electric birefringence measurements.
A delay line pulse generator, similar to that of Krause and
O'Konski, was used except that the cable was increased in length
to give a rectangular pulse of up to 8 kV amplitude and of 2.5
μsec duration. The relaxation time was obtained from the decay
curve, and the parameter r $(= \mu^{2}/[(\alpha_{1} - \alpha_{2})kT])$, which is the
ratio of the contributions of the permanent dipole moment and the
electrical anisotropy to the specific Kerr constant, was computed
from the ratio of the area bounded by the rise curve and that
bounded by the decay curve. As the pH was lowered from the
isoelectric point (pH 5.1), the relaxation time fell abruptly at
pH 4.1 and at pH 3.6. These decreases are consistent with the
two-stage expansion of the quadruple-unit molecule proposed by
Foster. The parameter r decreased continuously with decreasing
pH and the specific Kerr constant increased considerably below
pH 4. This suggests a large increase in the electrical anisotropy
of the BSA molecules in the low pH region. When the pH was

lowered below 3.2, the relaxation time increased very sharply.
This indicates the formation of large aggregates.

Next, they studied the transient electric birefringence of
solutions of ovalbumin, bovine γ-globulin, and β-lactoglobulin
[92]. Bovine γ-globulin and β-lactoglobulin were investigated
in aqueous solution, while ovalbumin was studied in both pure
water and in a glycerol-water mixture. For each protein, the
specific Kerr constant and the birefringence relaxation time were
measured over the concentration range of 0.3 - 1.7 g/100 ml, and
were extrapolated to infinite dilution. The relaxation time thus
obtained for ovalbumin and γ-globulin were compatible with
molecular models and dimensions presented in the literature. As
the pH was lowered from 6.7 (isoelectric point) to 3.0, the
relaxation time of γ-globulin in pure water, 0.004 M KCl, and
0.006 M KCl decreased gradually; the specific Kerr constant in
0.004 M KCl and 0.006 M KCl also decreased. This indicates some
changes in the molecular structure. The birefringence transients
of β-lactoglobulin upon application of a rectangular pulse were
unusual. The birefringence passed through a maximum, before
attaining the steady state. At the field cutoff, the birefringence dipped and then increased rapidly, before decaying in the
usual way. Tentatively, this unusual behavior was attributed to
the dissociation of aggregates, or of the parent molecule into its
subunits, under the influence of strong electric fields and the
recombination upon removal of the field.

G. Electric Birefringence of Hemoglobin

The three-dimensional structure of hemoglobin molecule in
the crystalline state has been revealed by x-ray investigations.
Orttung [93] obtained evidence for a permanent dipole moment along
the twofold axis in hemoglobin from the optical dispersion of
electric birefringence. The Kerr constant of aqueous solutions of
horse met- and oxyhemoglobin was measured at the three wavelengths

(356, 436, and 546 nm) in the visible absorption region. The
experimental results were expressed in terms of the Kerr constant
increment, k, defined by

$$B = B_0(1 + kc) \tag{5}$$

where B and B_0 are the Kerr constants of the solution and the
solvent, respectively, and c is the concentration in g/liter.
Using the molecular structure as determined in the crystalline
state, the intrinsic optical anisotropy of the four heme groups
per molecule was calculated as a function of wavelength for both
met- and oxyhemoglobin in two cases: (a) orientation of the
molecule with its long axis parallel to the field; and (b) orientation with the short or twofold axis parallel to the field.
Comparing the wavelength dependence of the experimental Kerr constant increment and that of the theoretical optical anisotropy, it
was concluded that the hemoglobin molecule had a permanent dipole
moment of about 100 D parallel to the twofold axis at neutral pH.

In a series of subsequent papers [94-98], Orttung gave a
detailed molecular interpretation of the dielectric properties and
the electric birefringence of hemoglobin. According to the
Kirkwood-Shumaker theory [99], proton fluctuations among the
available binding sites make a major contribution to the mean-
square dipole moment of proteins. First, this theory was extended
to include proton fluctuation anisotropy effects [94,95]. The
calculation of proton occupation averages at the individual binding sites was treated by an extension of the Tanford-Kirkwood
theory [100] of protein titration curves. Coordinates of the
proton-binding sites were obtained from the three-dimensional
structural models of the α and β chains of horse oxyhemoglobin,
and the theory was fitted to the experimental titration curve at
low ionic strength in the pH range 4.5-9.0 [96]. Then, the mean-
square dipole moment and the anisotropy of proton fluctuation
moment were calculated, using the parameters that gave the best
fit to the titration curve [97]. The mean-square dipole moment

was in good agreement with estimates from the dielectric increment data. The anisotropy of fluctuation moment was shown to be appreciable and to change with pH significantly.

Finally, the data for the Kerr effect optical dispersion of hemoglobin were interpreted in terms of dielectric and optical parameters [98]. The contribution of the heme spectrum to the anisotropy of optical polarizability was calculated from the Kramers-Kronig relation and geometrical information. The calculated Kerr constant increments were found to be very sensitive not only to the heme orientations and small x, y splittings of the Soret and visible bands, but also to the details of the anisotropy of proton fluctuation. The theory was in qualitative agreement with the experimental data.

H.   Electric Birefringence of Tobacco Mosaic Virus Proteins

Taniguchi et al. [101] studied the transient electric birefringence of monomers and oligomers of proteins from the ordinary strain of tobacco mosaic virus (TMV-0) and from the bean strain (TMV-B). These two proteins have a different amino acid composition. They exist as monomers $(M = 1.7 \times 10^4)$ in 33% acetic acid and as small oligomers (probably trimers) in 5mM phosphate buffer (pH 7.0). The specific Kerr constant (B/c) for the monomers of both proteins in 33% acetic acid was about 4×10^{-4} cm^4 statvolt^{-2} g^{-1}. The specific Kerr constant for the oligomer of TMV-0 protein in a 5 mM phosphate buffer was of the same order as for the monomers, while that for the oligomer of the TMV-B protein in the same solvent was two or three times larger. The birefringence relaxation time for the monomers of both proteins was 0.025-0.030 µsec in 33% acetic acid. In the presence of glycerol, the same value of the relaxation time was obtained after being corrected for the solvent viscosity. The relaxation time for the oligomers of both proteins was about 0.145 µsec. The parameter r was obtained from the rise curves; r = 2 for the

monomers of both proteins, $r = 0$ for the oligomer of TMV-0
protein, and $r = 1$ for the oligomer of TMV-B protein. These
results show that monomers are organized into oligomers in such a
way that the permanent dipole moment is cancelled in the case of
the TMV-0 protein, while it is not cancelled in the case of the
TMV-B protein. Thus, some qualitative difference between two
proteins in the process of constructing oligomers was suggested by
electric birefringence measurements.

REFERENCES

1. G. D. Fasman, ed., Poly-α-Amino Acids, Marcel Dekker, New
    York, 1967.

2. L. Pauling, R. B. Corey, and H. R. Branson, Proc. Nat. Acad.
    Sci. U.S., 37, 205 (1951).

3. P. Doty, J. H. Bradbury, and A. M. Holtzer, J. Am. Chem. Soc.,
    78, 947 (1956).

4. I. Tinoco, Jr., J. Am. Chem. Soc., 79, 4336 (1957).

5. C. T. O'Konski, K. Yoshioka, and W. H. Orttung, J. Phys.
    Chem., 63, 1558 (1959).

6. A. Wada, J. Chem. Phys., 30, 328, 329 (1959).

7. A. Wada, Bull. Chem. Soc. Japan, 33, 822 (1960).

8. K. Yamaoka, Ph.D. Thesis, University of California, Berkeley,
    1964.

9. G. Boeckel, J.-C. Genzling, G. Weill, and H. Benoit, J. Chim.
    Phys., 59, 999 (1962).

10. V. N. Tsvetkov, Yu. V. Mitin, V. R. Glushenkova, A. E.
    Grishchenko, N. N. Boitsova, and S. Ya. Lyubina,
    Vysokomolekul. Soedin., 5, 453 (1963).

11. V. N. Tsvetkov, I. N. Shtennikova, E. I. Ryumtsev, and
    V. S. Skazka, Vysokomolekul. Soedin., 7, 1111 (1965).

12. V. N. Tsvetkov, I. N. Shtennikova, V. S. Skazka, and E. I.
    Ryumtsev, J. Polymer Sci., Part C, 16, 3205 (1968).

13. O. Kratky and G. Porod, Rec. Trav. Chim., 68, 1106 (1949).

14. V. N. Tsvetkov, I. N. Shtennikova, E. I. Ryumtsev, and
    G. I. Okhrimenko, Vysokomolekul. Soedin., 7, 1104 (1965).

15. H. Watanabe, K. Yoshioka, and A. Wada, Biopolymers, 2, 91
    (1964).

16.   H. Watanabe and K. Yoshioka, Biopolymers, 4, 43 (1966).

17.   V. N. Tsvetkov, I. N. Shtennikova, E. I. Ryumtsev, and
      G. F. Pirogova, Vysokomolekul. Soedin., A9, 1583 (1967).

18.   V. N. Tsvetkov, I. N. Shtennikova, E. I. Ryumtsev, and
      G. F. Pirogova, Vysokomolekul. Soedin., A9, 1575 (1967).

19.   K. Nishinari and K. Yoshioka, Kolloid-Z. Z. Polymere, 235,
      1189 (1969).

20.   K. Nishinari and K. Yoshioka, Kolloid-Z. Z. Polymere, 240,
      831 (1970).

21.   K. Tsuji, H. Ohe, and H. Watanabe, Polym. J., 4, 553 (1973).

22.   H. Ohe, H. Watanabe, and K. Yoshioka, Colloid Polymer Sci.,
      262, 26 (1974).

23.   E. Marchal, C. Hornick, and H. Benoit, J. Chim. Phys., 64,
      514 (1967).

24.   M. Matsumoto, H. Watanabe, and K. Yoshioka, Biopolymers, 6,
      929 (1968).

25.   M. Matsumoto, H. Watanabe, and K. Yoshioka, Kolloid-Z. Z.
      Polymere, 250, 298 (1972).

26.   M. Matsumoto, H. Watanabe, and K. Yoshioka, Biopolymers, 9,
      1307 (1970).

27.   M. Matsumoto, H. Watanabe, and K. Yoshioka, Biopolymers, 12,
      1729 (1973).

28.   J. C. Powers, Jr., J. Am. Chem. Soc., 88, 3679 (1966).

29.   J. C. Powers, Jr., J. Am. Chem. Soc., 89, 1780 (1967).

30.   R. F. Epand and H. A. Scheraga, Biopolymers, 6, 1383 (1968).

31.   F. Joubert, N. Lotan, and H. A. Scheraga, Physiol. Chem.
      Phys., 1, 348 (1969).

32.   K. Kikuchi and K. Yoshioka, Biopolymers, 12, 2667 (1973).

33.   K. Kikuchi and K. Yoshioka, J. Phys. Chem., 77, 2101 (1973).

34.   E. Neumann and A. Katchalsky, Proc. Nat. Acad. Sci. U.S.,
      69, 993 (1972).

35.   Y. Okamoto and R. Sakamoto, Nippon Kagaku Zasshi, 90, 669
      (1969).

36.   R. Sakamoto and Y. Okamoto, Nippon Kagaku Zasshi, 90, 1780
      (1967).

37.   H. Watanabe, Nippon Kagaku Zasshi, 86, 179 (1965).

38.   J. C. Powers, Jr. and W. L. Peticolas, Adv. Chem. Ser., 63,
      217 (1967).

39.   J. C. Powers, Jr. and W. L. Peticolas, Biopolymers, 9, 195
      (1970).

40. M. Matsumoto, H. Watanabe, and K. Yoshioka, Biopolymers, 11, 1711 (1972).

41. B. L. Brown and B. R. Jennings, Biopolymers, 9, 1119 (1970).

42. K. Yoshioka and H. Watanabe, Nippon Kagaku Zasshi, 84, 626 (1963); K. Yoshioka and H. Watanabe, in Physical Principles and Techniques of Protein Chemistry, Part A (S. J. Leach, ed.), Academic, New York, 1969, p. 335.

43. G. Spach, Compt. Rend. Acad. Sci. (Paris), 249, 667 (1959).

44. T. Miyazawa, in Poly-α-Amino Acids (G. D. Fasman, ed.), Marcel Dekker, New York, 1967, p. 96.

45. A. Maschka, G. Bauer, and Z. Dora, Mh. Chem., 101, 1516 (1971).

46. J. V. Champion and J. W. Nicholas, Chem. Phys. Lett., 14, 573 (1972).

47. M. Shirai, Abstracts, 152nd National Meeting of the American Chemical Society, New York, Sept. 1966, p. 131.

48. E. Charney, J. B. Milstien, and K. Yamaoka, J. Am. Chem. Soc., 92, 2657 (1970).

49. J. B. Milstien and E. Charney, Biopolymers, 9, 991 (1970).

50. T. C. Troxell and H. A. Scheraga, Biochem. Biophys. Res. Commun., 35, 913 (1969).

51. T. C. Troxell and H. A. Scheraga, Macromolecules, 4, 519 (1971).

52. T. C. Troxell and H. A. Scheraga, Macromolecules, 4, 528 (1971).

53. B. R. Jennings and E. D. Baily, Nature [Phys. Sci.], 233, 162 (1971).

54. E. D. Baily and B. R. Jennings, Appl. Opt., 11, 527 (1972).

55. E. Iizuka, Biochim. Biophys. Acta, 175, 457 (1969).

56. E. Iizuka, Biochim. Biophys. Acta, 243, 1 (1971).

57. E. Iizuka, Polymer J., 5, 62 (1973).

58. C. Wippler, J. Chim. Phys., 53, 346 (1956).

59. C. Wippler, J. Polymer Sci., 23, 199 (1957).

60. M. L. Wallach and H. Benoit, J. Polymer Sci., 57, 41 (1962).

61. B. R. Jennings and H. G. Jerrard, J. Phys. Chem., 69, 2817 (1965).

62. Y. Mukohata, S. Ikeda, and T. Isemura, J. Mol. Biol., 5, 570 (1962).

63. Y. Mukohata, Ann. Rep. Sci. Works, Fac. Sci., Osaka Univ., 11, 1 (1963).

64. S. Ikeda, J. Chem. Phys., 38, 2839 (1963).

65. Y. Mukohata, J. Mol. Biol., 7, 442 (1963).

66. V. N. Tsvetkov and E. L. Vinogradov, Vysokomolekul. Soedin., 8, 662 (1966).

67. I. Tinoco, Jr. and J. D. Ferry, J. Am. Chem. Soc., 76, 5573 (1954).

68. I. Tinoco, Jr., J. Am. Chem. Soc., 77, 3476 (1955).

69. I. H. Billick and J. D. Ferry, J. Am. Chem. Soc., 78, 933 (1956).

70. A. E. V. Haschemeyer and I. Tinoco, Jr., Biochemistry, 1, 996 (1962).

71. D. N. Holcomb and I. Tinoco, Jr., J. Phys. Chem., 67, 2691 (1963).

72. A. E. V. Haschemeyer, Biochemistry, 2, 851 (1963).

73. K. Yoshioka and C. T. O'Konski, Biopolymers, 4, 499 (1966).

74. L. D. Kahn and L. P. Witnauer, J. Am. Leather Chem. Soc., 64, 12 (1969).

75. L. D. Kahn and L. P. Witnauer, Biochim. Biophys. Acta, 243, 388 (1971).

76. S. Ananthanarayanan and A. Veis, Biopolymers, 11, 1365 (1972).

77. J. C. Bernengo, B. Roux, and D. Herbage, Biopolymers, 13, 641 (1974).

78. H. Asai, J. Biochem. (Tokyo), 50, 182 (1961).

79. S. Kobayasi, H. Asai, and F. Oosawa, Biochim. Biophys. Acta, 88, 528 (1964).

80. S. Kobayasi, Biochim. Biophys. Acta, 88, 541 (1964).

81. R. M. Pytkowicz and C. T. O'Konski, Biochim. Biophys. Acta, 36, 466 (1959).

82. D. Ridgeway, J. Am. Chem. Soc., 90, 18 (1968).

83. D. Ridgeway, J. Am. Chem. Soc., 88, 1104 (1966).

84. J. E. Mellema and A. Klug, Nature, 239, 146 (1972).

85. S. Krause, Ph.D. Thesis, University of California, Berkeley, 1957.

86. S. Krause and C. T. O'Konski, J. Am. Chem. Soc., 81, 5082 (1959).

87. S. Krause and C. T. O'Konski, Biopolymers, 1, 503 (1963).

88. P. Moser, P. G. Squire, and C. T. O'Konski, J. Phys. Chem., 70, 744 (1966); P. Moser, P. G. Squire, and C. T. O'Konski, Biochemistry, 7, 4261 (1968).

89. P. Ingram and H. G. Jerrard, Nature, 196, 57 (1962).

90. H. G. Jerrard, C. L. Riddiford, and P. Ingram, J. Phys. E: Sci. Instr., 2, 761 (1969).

91. C. L. Riddiford and B. R. Jennings, J. Am. Chem. Soc., 88, 4359 (1966).

92. C. L. Riddiford and B. R. Jennings, Biopolymers, 5, 757 (1967).

93. W. H. Orttung, J. Am. Chem. Soc., 87, 924 (1965).

94. W. H. Orttung, J. Phys. Chem., 72, 4058 (1968).

95. W. H. Orttung, J. Phys. Chem., 72, 4066 (1968).

96. W. H. Orttung, J. Am. Chem. Soc., 91, 162 (1969).

97. W. H. Orttung, J. Phys. Chem., 73, 418 (1969).

98. W. H. Orttung, J. Phys. Chem., 73, 2908 (1969).

99. J. G. Kirkwood and J. B. Shumaker, Proc. Nat. Acad. Sci. U.S., 38, 855 (1952).

100. C. Tanford and J. G. Kirkwood, J. Am. Chem. Soc., 79, 5333 (1957).

101. M. Taniguchi, S. Kobayashi, and A. Yamaguchi, Biochim. Biophys. Acta, 188, 140 (1969).

Chapter 18

ELECTRO-OPTICS OF POLYNUCLEOTIDES AND NUCLEIC ACIDS

Nancy C. Stellwagen

1409 Cedar Street
Iowa City, Iowa

I.   INTRODUCTION

Nucleic acids (naturally occurring polynucleotides) and
synthetic polynucleotides are high molecular weight ribose- or
deoxyribose-phosphates to which purine and/or pyrimidine bases are
attached.   The purines are usually adenine and guanine; the
pyrimidines, thymine, cytosine, and uracil.   Deoxyribonucleic acid
(DNA) has been shown by Watson and Crick [1] and by Wilkins and
coworkers [2,3] to be a double-stranded helix with adenine-thymine

and guanine-cytosine base pairs oriented perpendicular to the
helix axis. X-ray studies [4] of DNA solutions, low-angle x-ray
scattering measurements [5] and studies of light scattering at
large angles [6], have shown that this structure persists in
solution. However, the conformation of the DNA molecule in solution has not been completely established. At low molecular
weights ($\leq$ ca. 200,000) the DNA molecule appears to be relatively rigid [7-11]. At higher molecular weights coiling sets in [7-
10], usually described in terms of the Kratky-Porod wormlike chain
[12] or the bead subchain model of Zimm-Rouse-Bueche [13-15]. A
rigid zigzag chain model has also been proposed [6]. The structures and conformations of the ribonucleic acids (RNA) are not as
well known as those of DNA. Hydrodynamic [16,17], electric
dichroism [18-20], electron microscope [21,22], and low-angle
x-ray [23,24] studies of high molecular weight RNA molecules in
solution suggest they contain short intramolecular helical segments separated by flexible single-stranded sections. Low molecular weight (transfer) RNA molecules are believed to have a clover-
leaf structure containing about four helical regions [25-27].
Synthetic polynucleotides in solution can assume the structure of
double- or triple-stranded helices, single-stranded helices
stabilized by base stacking, or random coils, depending on experimental conditions such as pH, temperature, and ionic strength [28,
29].

In this chapter the results obtained from electro-optic
studies of polynucleotides and nucleic acids will be summarized.
It will be seen that electric birefringence and other electro-
optic techniques can be used to deduce the conformation of
polynucleotides in solution, to study reaction kinetics and the
binding of small ions to polyions, and to elucidate the electrical
and optical properties of polynucleotides.

II.   DEOXYRIBONUCLEIC ACID

Deoxyribonucleic acids from different animal and bacterial sources have been studied extensively, but the results are not always directly comparable because of differences in the molecular weights of the samples, DNA concentration, and electrolyte concentration.

A. Sign of the Birefringence and Dichroism

DNA molecules in solution exhibit negative electric birefringence, as pointed out by Benoit [30], Haltner and O'Konski [31,32], and others [33-39]. Negative flow birefringence has been observed by Signer et al. [40], Snellman and Widström [41], and others [42,43,43a]. Negative birefringence in DNA fibers was noted by Astbury and Bell [44] and by Seeds and Wilkins [45]; the latter also noted perpendicular dichroism in the fibers. Dvorkin [46-48] and Ding et al. [49] observed that DNA molecules in solution exhibit perpendicular electric dichroism. Both negative birefringence and perpendicular dichroism indicate that the direction of maximum polarizability is perpendicular to the long axis of the molecule. Since the purine and pyrimidine bases have high optical polarizabilities in the planes of the molecules, negative birefringence and perpendicular dichroism are consistent with the orientation of the base pairs perpendicular to the long axis of the Watson-Crick double helix.

When the ordered structure of DNA is disrupted by heating above the melting temperature [7,8] of the molecules, the dichroism disappears [46] and a small positive birefringence signal is observed [33,43a,50]. These results indicate that the purine and

pyrimidine bases have a random orientation with respect to the
long axis of the molecules. Positive birefringence is only observed at elevated temperatures [33,50]; solutions which have been
heated above the melting temperature and cooled to room temperature exhibit negative birefringence [50,51]. If formaldehyde is
present in the solution to prevent renaturation of the separated
DNA strands, positive birefringence can be observed in solutions
cooled to room temperature [50]. The positive birefringence of
the separated DNA strands is characterized by very long relaxation
times, of the order of milliseconds [50], indicating that the
separated strands are very extended and rotate as relatively rigid
units in the electric field [50].

     The electric dichroism [46,47] and flow dichroism [52] exhi-
bited by DNA molecules in solution decrease markedly at pH $\leq$ 3,
corresponding to the region of acid denaturation of DNA [7,8].
The dichroic ratio (see Chap. 7) increases at pH 6.5 and remains
at this higher value to pH 10 [46,47]. The higher dichroic ratio
observed at basic pH values is consistent with viscosity and flow
dichroism measurements indicating a maximum stability of the DNA
helix at pH 8-9 [52]. The dichroic ratio of DNA in 80% ethanol is
lower than in aqueous solutions [49], suggesting that a structural
change may occur in this solvent [53].

B.   Relaxation Behavior

     A typical oscilloscope trace of the electric birefringence of
calf thymus DNA in 10^{-4} M Tris buffer, pH 8, is shown in Fig. 1.
It is apparent that a wide range of relaxation times is present in
this sample. A graphical analysis of the decay of the birefringence is presented in Fig. 2. The longest relaxation time,
τ_{long} = 105 msec, was obtained from the limiting slope of the
normalized experimental curve. Shorter relaxation times were then
obtained by subtracting the contribution of τ_{long} from the
experimental curve and continuing the analysis in the same manner
([54]; see also Chap. 3).

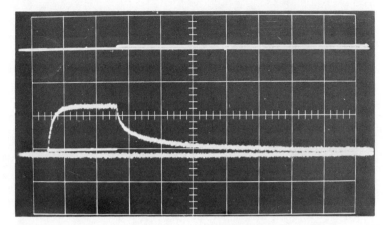

FIG. 1.   Oscilloscope trace of electric birefringence signal of calf thymus DNA in 10^{-4} M Tris buffer, pH 8.   The applied pulse is also shown (inverted).   DNA concentration = 0.0076 g/liter, E = 920 V/cm.   One horizontal scale division = 2 msec.

A summary of the birefringence relaxation times observed by different investigators studying various DNA samples is given in Table 1.   Selected data obtained by other techniques are also included.   Where necessary, rotational diffusion constants given by some authors have been converted to birefringence relaxation times in order to have a common basis of comparison.   Although most investigators observed that the spectrum of birefringence relaxation times was not unimodal, only the longest relaxation times are listed in Table 1.   The data are arranged approximately in order of increasing relaxation time.   A dashed line indicates that the information was not available.

Table 1 illustrates that the relaxation times reported by different investigators range from 0.20 to 570 msec.   This variation is larger than one might expect from differences in molecular weight and experimental conditions.   Some of the smaller relaxation times are due to earlier workers who used partially denatured samples or samples of low molecular weight.   The presence of residual protein in the samples may also affect the results;

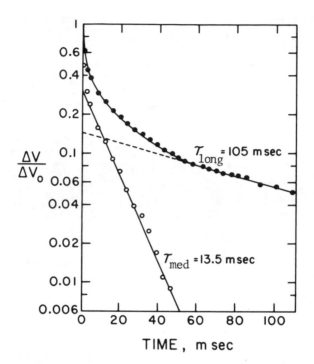

FIG. 2. Analysis of birefringence decay curve of calf thymus
DNA in 10^{-4} M Tris buffer. DNA concentration = 0.0071 g/liter,
E = 1150 V/cm. The logarithm of the birefringence, ΔV, relative
to the steady state birefringence, ΔV_0, is plotted versus time
after removal of the electric field. (●) experimental points; (○)
with the long component subtracted out. Still shorter relaxation
times could be obtained from an analysis of the decay curve on an
expanded time scale (not shown).

Reinert et al. [73] found that a finite concentration of residual
protein increased the rigidity of calf thymus DNA molecules, and
Thurston and Wilkinson [43a] found that salmon sperm DNA exhibited
shorter relaxation times after extraction with phenol to remove
bound protein. However O'Konski and Stellwagen [37,51] found that
calf thymus DNA samples from five different sources exhibited very
similar birefringence relaxation times. Differences in DNA concentration may also affect the observed relaxation times; many of the
shorter relaxation times in Table 1 ($\leq$ 15 msec) were obtained
with relatively concentrated solutions. It is possible that DNA

TABLE 1

Maximum Relaxation Times Observed for DNA Molecules in Solution

Investigator	Ref.	DNA source	DNA conc. (g/liter)	Solvent medium	τ_{max} (msec)	Technique
Ingram and Jerrard	55,56	Calf thymus	0.1	Water	0.20	Electric birefringence
Jennings and Plummer	57	Calf thymus	--	Water	0.39	Electric field light scattering
Golub et al.	33,34,58	Calf thymus	0.02	2×10^{-3} M NaCl	0.53	Electric birefringence
		T2 phage	0.02	2×10^{-3} M NaCl	0.47	Electric birefringence
		Calf thymus	0.02	Water	9.2	Electric birefringence
Benoit	30	Calf thymus	0.01	Water	2.5	Electric birefringence
Allais	59	Cancer cells	0.1	Water-acetone-ether	2.5	Electric birefringence
Ding et al.	49	Calf thymus	0.02-0.05	10^{-4} M Tris	3.0	Electric dichroism
Weill et al.	11	Calf thymus	--	5×10^{-3} M NaCl-citrate	2.5	Electric birefringence
Hornick and Weill	60	--	0.1	5×10^{-3} M NaCl	6.0	Electric birefringence
Cavalieri et al.	61	Calf thymus	0.2-1.0	0.2 M NaCl	3.7	Flow dichroism
				10^{-3} M NaCl	9.1	Flow dichroism

TABLE 1 (continued)

Investigator	Ref.	DNA source	DNA conc. (g/liter)	Solvent medium	τ_{max} (msec)	Technique
Schwander and Cerf	62	Calf thymus	0.06–0.2	Water	13	Flow birefringence
Mathieson and Matty	63	Calf thymus	0.15	0.1 M NaCl	15.6	Flow birefringence
Ise et al.	64	Calf thymus	0.016	Water	100	Anisotropy of conductivity
Frisman et al.	10	Calf thymus	0.1	0.15 M NaCl	100	Flow birefringence
Goldstein and Reichman	65	Calf thymus	0.01–0.06	0.1 M NaCl	101	Flow birefringence
O'Konski et al.	37,51	Calf thymus	0.007	10^{-4} M Tris	105	Electric birefringence
				10^{-3} M Tris	18	Electric birefringence

Itzhaki	66	Rat thymus	0.04	10^{-4} M phosphate buffer	180	Electric birefringence
Ohlenbusch	67	--	--	--	200	Electric birefringence
Thurston and Wilkinson	43a	Salmon sperm	3.0	Deionized water	200	Oscillatory flow birefringence
				0.1 M NaCl	90	Oscillatory flow birefringence
Harrington	68	T2 phage	0.005–0.04	0.16 M NaCl-citrate	300	Flow birefringence
Callis and Davidson	69	T4 phage	0.008	0.1 M NaCl	450	Decay of flow dichroism
Thompson and Gill	70	T2 phage	0.029	0.2 M BPES buffer [9]	450–520	Decay of flow birefringence
Chapman and Hall	71	T2 phage	0.01–0.03	Glycerol, BBES buffer [72]	570	Creep recovery

molecules in more concentrated solutions have a relaxation
mechanism associated with bending motions, or motions of segments,
while more dilute solutions exhibit a combination of bending
motions and whole molecule rotations. Flow birefringence and
viscosity studies [74-78] have also been interpreted in terms of
segmental orientation of the DNA molecules.

Relaxation times of 100 msec and over in Table 1 probably
correspond to rotation of the DNA molecule as a whole. The relaxation times of T2 and T4 phage DNA (300-570 msec) are larger than
those of calf thymus DNA (100 msec), reflecting the higher molecular weight of the phage DNA [7,79]. However, lengths calculated
from these relaxation times are smaller than would be obtained if·
the molecules rotated as rigid rods in solution. For example, the
length calculated for calf thymus DNA in 10^{-4} M Tris buffer, using
the Burgers-Broersma equation [80] and a diameter of 26 Å [8] or
60 Å [4], was found to be 25,000 Å, about half the estimated contour length of 40,000 Å [37,51]. Hence the DNA molecules must
have been somewhat coiled in this solvent. Hydrodynamic studies
[7-9,73,81-86] indicate that the DNA molecule is best described as
a coil of continuous limited flexibility; i.e., a wormlike coil
[12]. Electron microscope studies [87] also suggest that the
conformation of the DNA molecule in solution is intermediate
between that of a rigid rod and a random coil.

Table 1 also illustrates that the relaxation times decrease
with increasing ionic strength of the solvent [33,34,37,43a,51,58,
61]. For example, the relaxation time of the DNA sample studied
by O'Konski et al. [37,51] decreased from 105 to 18 msec when the
Tris buffer concentration was increased from 10^{-4} to 10^{-3} M.
Using the Burgers-Broersma equation [80] and a diameter of 26 Å
[8], this decrease corresponds to a decrease in length from 25,000
to 14,000 Å [51], and can be attributed to the greater coiling of
the DNA molecule in the solution of higher buffer concentration.
Hydrodynamic [7,8,82,85,86] and electron microscope [88] studies
also indicate that DNA molecules are more highly coiled in
solutions of higher ionic strength.

Stellwagen [37] analyzed the fast portion of the decay of the birefringence of calf thymus DNA by finding the time required for a line tangent to the initial portion of the decay curve to inter-sect the axis of zero birefringence. This time interval is de-fined as t_0 [54]. She found that the t_0 values at a given electric field strength were essentially independent of ionic strength, as shown in Fig. 3, in contrast to the marked dependence of τ_{max} on buffer concentration (see Table 1 and related discus-sion). The t_0 values decrease with increasing electric field strength, probably due to a saturation effect [90]. The constancy of the t_0 values at a given field strength suggests that the DNA molecule orients by segments in the electric field [37], and that the size of the segment is independent of buffer concentration, at least in the range studied (10^{-5} M to 2×10^{-3} M). The average

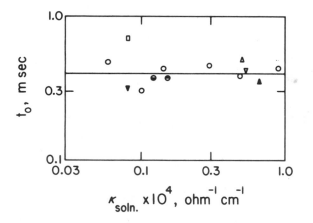

FIG. 3. Initial decay of the birefringence, characterized by t_0 (see text for definition), as a function of the conductivity of the solution. E ≃ 1000 V/cm. (o) DNA concentration = 0.007 g/liter in Tris buffer solutions ranging from 10^{-5} to 2×10^{-3} M; (△) 0.0035 g DNA/liter in 10^{-3} M Tris; (▽) 0.014 g DNA/liter in 10^{-3} M Tris; (□) 0.0035 g DNA/liter in 10^{-4} M Tris; (▲) 0.007 g DNA/liter in 10^{-3} M cacodylate buffer; (▼) 0.007 g DNA/liter in 10^{-5} M MgCl$_2$; (◉) 0.007 g DNA/liter in 10^{-4} M Tris + 10^{-5} M MgCl$_2$; (◓) 0.007 g DNA/liter in 10^{-4} M Tris + 10^{-5} M AgNO$_3$.

t_0 value observed at $E \simeq 1,000$ V/cm was 0.4 msec, corresponding
to a segment length of 3,500 Å, using the Burgers-Broersma equation [80] and a diameter of 26 Å [8].

Ding et al. [49] analyzed the decay of the electric dichroism
of calf thymus DNA and sonicated fragments in high electric fields
using the method of Isenberg and Dyson [89]. The dichroism decay
curves contained several components, including a slow one dependent on molecular weight and a fast component essentially independent of molecular weight. The fast component was attributed to
rotation of segments of the DNA molecule. From the average
relaxation time of 8 μsec, a segment length of 800 Å was calculated [49] using the Burgers-Broersma equation [80] and a diameter
of 26 Å [8]. Ding et al. [49] suggest that a segment of this
length represents the smallest unit capable of independent rotation in an electric field.

In very high electric fields ($\geq$ 10 kV/cm) O'Konski and
Stellwagen [91] found that the birefringence signal of calf thymus
DNA in 10^{-4} M Tris buffer reaches a maximum and begins to decrease
while the pulse is still being applied to the solution. The decay
of the birefringence also exhibits an anomaly [91]. Ding et al.
[49] found that the electric dichroism signal of calf thymus DNA
in 10^{-4} M Tris buffer also goes through a maximum in very high
electric fields before the pulse is terminated. However, Yamaoka
and Charney [91a] did not observe similar effects at high electric
fields with calf thymus DNA in 10^{-3} M NaCl. The maximum observed
in the birefringence and dichroism signals at high electric
fields can be attributed to the simultaneous presence of both
positive and negative components in the signal [91], possibly due
to aggregation of the DNA molecules and a tilting of the base
pairs in very high electric fields [91]. Alternatively, a partial
unwinding of the DNA helix may occur in high electric fields.
Ribosomal RNA and a synthetic polynucleotide, poly (A + 2U), have
been observed to partially unwind in very high electric fields
[92]. Implications of these effects are discussed more completely
in Chap. 3, Sec. IVD.

An analysis of the birefringence decay curve of a sample of sonicated calf thymus DNA is shown in Fig. 4. At low field strengths the decay of the birefringence was unimodal, with a relaxation time of 49 μsec. At higher field strengths, a second component with a relaxation time of 11 μsec was also observed [37]. Weill et al. [11,60], Ding et al. [49], and Thurston and Wilkinson [43a] have also observed that the relaxation times of sonicated DNA particles are much smaller than those of the parent molecules. However, Golub [33,34] found essentially no change in the relaxation times observed when T2 phage molecules were sonicated, probably because of the very low relaxation times observed for the intact molecules (see Table 1). Representative relaxation times observed for sonicated DNA particles by different investigators are given in Table 2. Where more than one relaxation time was

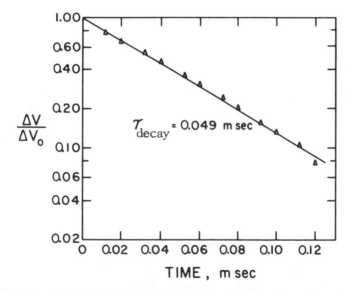

FIG. 4. Analysis of birefringence decay curve of sonicated calf thymus DNA in 10^{-3} M Tris buffer. Concentration of sonicate = 0.0076 g/liter, E = 1200 V/cm. The logarithm of the birefringence, ΔV, relative to the steady state birefringence, ΔV_0, is plotted against time after the removal of the electric field.

TABLE 2

Relaxation Times and Calculated Lengths of Sonicated DNA Particles

Investigator	Ref.	DNA source	Mol. wt. $\times 10^5$	Sonicate conc. (g/liter)	Buffer conc.	τ_{max} (μsec)	Length (Å)
O'Konski et al.	51	Calf thymus	--	0.007	10^{-3} M Tris	49	1630
Ding et al.	49	Calf thymus	3.6	0.02–0.04	2×10^{-4} M cacodylic acid	76	1900
Hornick and Weill	60	Calf thymus	3.9	0.1	5×10^{-3} M NaCl–citrate	107	2200
Thurston and Wilkinson	43a	Salmon sperm	6	3	Deionized water	500	4100
Golub	33, 34	T2 phage	--	0.02	10^{-3} M NaCl–citrate	500	4100

observed, the longer one is given. Particle lengths, calculated
from the Burgers-Broersma [80] or Burgers [93] equations are also
given. The diameter assumed in most cases was 26 Å.

Ding et al. [49] analyzed the relaxation times of a series of
sonicated or otherwise cleaved fragments of calf thymus DNA. The
data for low molecular weight samples ($M \leq 3.6 \times 10^5$) appeared
to fit the Burgers-Broersma [80] equation for stiff rods, while
high molecular weight samples were better described by the Zimm-
Rouse-Bueche [13-15] theory. The transition appeared to occur at
about 5 persistence lengths [49].

C. Apparent Kerr Constants

In order to compare the magnitude of the birefringence
observed for various DNA samples by different investigators,
apparent Kerr constants, K_{app}, have been calculated from the
data in the literature. Since the Kerr law [94] is generally not
obeyed by DNA molecules (see following section), the apparent Kerr
constant will depend on the magnitude of the electric field (E)
at which the measurement was made. The values of K_{app} and E,
where known, are summarized in Table 3.

Two conclusions can be drawn from Table 3: the apparent Kerr
constant of DNA decreases as the ionic strength of the solution is
increased, and sonicated DNA particles have smaller apparent Kerr
constants than intact molecules. The K_{app} values in the first
and second halves of the table are reasonably self-consistent,
although they do not agree. For example, the average value of
K_{app} observed for whole DNA molecules in 10^{-3} M buffer by three
different investigators in three different laboratories was
$(-6.5 \pm 0.9) \times 10^{-3}$ cgs esu [33,35,37]. Two groups of investigators in another laboratory [11,95,96] observed $K_{app} =$
$(-0.76 \pm 0.05) \times 10^{-3}$ cgs esu. The latter investigators may have
obtained their data at higher electric field strengths, which

TABLE 3

Apparent Kerr Constants[a] of DNA

Investigator	Ref.	DNA source	Buffer conc.	E (V/cm)	$-K_{app}^{a} \times 10^{3}$, (cgs esu)
Stellwagen	37	Calf thymus	10^{-3} M Tris	200	6.2
			10^{-4} M Tris	200	11.4
		Sonicate	10^{-3} M Tris	1000	0.41
Golub	34	Calf thymus, T2 phage	10^{-3} M NaCl-citrate	--	7.8
Houssier and Fredericq	35	Calf thymus	10^{-3} M NaCl	--	5.4
Itzhaki	66	Rat thymus	10^{-4} M phosphate buffer	250	16.2
Benoit	95,96	Calf thymus	Water	--	0.71
Weill et al.	11	Calf thymus	10^{-3} M NaCl	--	0.81
		Sonicate	10^{-3} M NaCl	--	0.12
Hornick and Weill	60	Sonicate	10^{-3} M NaCl	--	0.082

[a]Calculated as though the Kerr law were obeyed.

would lead to lower apparent Kerr constants since the Kerr law is
not obeyed (see below).

D. Saturation of the Birefringence and Dichroism

At low field strengths DNA molecules in solution do not obey
the Kerr law; i.e., the magnitude of the birefringence or dichroism is not proportional to the square of the electric field
strength [35,37,66,67,91a,97]. Sonicated DNA particles [37],
ribosomal RNA molecules [50,98], gel-forming deoxyribonucleohis-
tones [99], sodium polyphosphates [37], and potassium
polystyrenesulfonate in water-glycerol and ethylene glycol mixtures [100,101] also do not obey the Kerr law at low field
strengths. Several investigators [35,60,91a,97,98] have noted an
approximately linear dependence of the birefringence or dichroism
of DNA and RNA on electric field strength. Stellwagen [37] observed that the birefringence of calf thymus DNA in 10^{-3} or 10^{-4} M
Tris buffer is proportional to $E^{0.8}$, as shown in Fig. 5. This
proportionality extends over a thirtyfold variation in electric
field strength. Figure 5 also illustrates that the smaller birefringence observed in solutions of higher buffer concentration is
due to a decrease in the electrical polarizability of the molecules, as predicted from an ion atmosphere polarization mechanism
[90,102-105]. At very high field strengths both curves in Fig. 5
converge toward the same limit, as expected since the magnitude of
the birefringence at complete orientation is determined solely by
the optical anisotropy of the molecules [90], which is the same at
both buffer concentrations.

Yamaoka and Charney [91a] were unable to reach saturation of
the electric dichroism of calf thymus DNA in 10^{-3} M NaCl using
electric fields up to 8 kV/cm. Ding et al. [49], using higher
electric fields, found that the saturation of the electric dichroism of calf thymus DNA in a Tris-cacodylate buffer could be described by the saturation equations of O'Konski et al. [90] for the

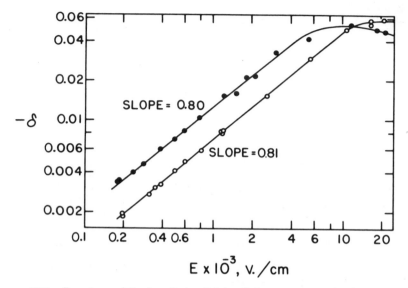

FIG. 5. Logarithmic plot of birefringence saturation curves of calf thymus DNA in 10^{-3} and 10^{-4} M Tris buffer. DNA concentration = 0.0074 g/liter. The retardation, δ, which is proportional to the birefringence, Δn, is plotted versus electric field strength. The Kerr law predicts a slope of 2.0 at low field strengths.

case of permanent moment orientation. The limiting values of $\Delta\varepsilon/\varepsilon$ (see Chap. 7) obtained from extrapolation of the theoretical curves ranged from -1.30 to -1.35, almost the same as the value of -1.31 calculated from flow dichroism data [105]. This value of $\Delta\varepsilon/\varepsilon$ corresponds to an angle of about 80° between the transition moment and the dipole axis [49]. Another polyelectrolyte, sodium polyethylenesulfonate, also exhibits birefringence saturation behavior characteristic of permanent moment orientation [90]. Neumann and Katchalsky [92] have pointed out that the polarization of bound counterions of rigid rod-like polyelectrolytes should approach saturation at moderately low field strengths. Therefore, in high electric fields large rigid polyelectrolytes should behave as though they possess a nearly constant dipole moment [49]. A stochastic configurational dipole moment has also been proposed to explain the apparent dipole moments of polyelectrolytes [51,90].

E.   Optical Anisotropy Factors

Optical anisotropy factors, $(g_1 - g_2)$, which have been obtained for DNA and sonicated DNA particles by various techniques are summarized in Table 4. The optical factors calculated from flow birefringence measurements are an order of magnitude lower than those obtained from electric birefringence saturation [37] or electric field light scattering [11] studies, even on the same sample [11]. The reason for this discrepancy is not clear. From flow birefringence measurements Goldstein and Reichman [65] concluded that $(g_1 - g_2)$ decreased with increasing molecular weight, while Tsvetkov, Frisman, and coworkers [10,42,74] found that $(g_1 - g_2)$ increased with increasing molecular weight. Stellwagen [37], using the electric birefringence saturation technique [90], found that the optical factor was independent of molecular weight, as predicted by Stuart and Peterlin [106] and O'Konski et al. [90].

The optical factor of DNA can also be calculated from the birefringence of highly oriented fibers. From the value of $\Delta n = -0.11$ reported by Seeds and Wilkins [45], the optical factor can be calculated to be -2.3×10^{-2} [37], in agreement with the values obtained from electric birefringence saturation [37] and electric field light scattering [11] measurements.

F.   Electrical Anisotropy and the Question of the Dipole Moment

The electrical anisotropy of DNA can be expressed in terms of the dipole moment, μ, and/or the electrical polarizability, $(\alpha_1 - \alpha_2)$. Relevant data from the literature are collected in Table 5. At present, there is no agreement about the possible existence of a permanent dipole moment in DNA molecules. Jennings and Plummer [57], using electric field light scattering, found a dipole moment of 7.4×10^4 D directed along the long axis of the molecule, while Scheludko and Stoylov [107], using the same

TABLE 4

Optical Anisotropy Factors of DNA

Investigator	Ref.	DNA source	DNA conc. (g/liter)	Solvent	$-(g_1 - g_2) \times 10^2$	Technique
Stellwagen	37	Calf thymus	0.0074	10^{-3} M Tris	2.5	Electric birefringence saturation
		Sonicate	0.0076	10^{-3} M Tris	2.4	Electric birefringence saturation
Weill et al.	11	Sonicate	0.1	0.2 M NaCl-citrate	3.4	Electric field light scattering
					0.07	Flow birefringence
Hornick and Weill	60	Sonicate	0.1	0.005 M NaCl	0.68	Flow birefringence
Goldstein and Reichman	65	Calf thymus	0.01–0.06	0.1 M NaCl	0.10	Flow birefringence
Frisman et al.	42	Calf thymus	0.1	0.15 M NaCl	0.14	Flow birefringence

TABLE 5

Electrical Anisotropy of DNA

Investigator	Ref.	DNA source	DNA conc. (g/liter)	Solvent	$\mu \times 10^{-4}$ (D)	$(\alpha_1 - \alpha_2) \times 10^{14}$ (cm^3)	Technique
Jennings and Plummer	57	Calf thymus	0.76	Water	7.4	13	Electric field light scattering
Scheludko and Stoylov	107	Calf thymus	0.02	10^{-4} M KCl	0.10 ($\perp$)	80	Electric field light scattering
Ise et al.	64	Calf thymus	0.01-0.1	Water	--	0.91	Conductivity dispersion
Dvorkin and Golub	33	Calf thymus	0.02	2×10^{-3} M NaCl	--	2.3	Electric birefringence
Hornick and Weill	60	Sonicate	0.1	10^{-3} M NaCl	0	0.14	Flow and electric birefringence
				Water	0	2.22	Flow and electric birefringence
			0.41	10^{-3} M NaCl	0	0.12	Electric field light scattering
			0.1	10^{-3} M NaCl	0	0.10	Electric dichroism

technique, interpreted their data as indicating a permanent dipole moment of 10^3 D directed along the transverse axis. The latter interpretation has since been questioned by Stoylov [108]. Benoit [96] and Hornick and Weill [60] found no evidence of permanent dipole orientation in the buildup of the electric birefringence, and therefore interpreted their results in terms of electric polarizability. Ding et al. [49] found that the buildup of the electric dichroism was characteristic of permanent moment orientation and the rate of approach of the dichroism to saturation could be described by saturation curves calculated for permanent moment orientation [90]. Electric birefringence saturation methods are ambiguous because the Kerr law is not obeyed [37]. The possible physical basis of a dipole moment in large rigid polyelectrolytes has been discussed by O'Konski et al. [51] and by Neumann and Katchalsky [92].

Hornick and Weill [60] obtained consistent results for the electrical polarizability of sonicated DNA particles using several different techniques: electric and flow birefringence (to eliminate the optical factor), electric field light scattering, electric field dichroism (using proflavine-DNA complexes), and electric field fluorescence (using acridine orange-DNA complexes [109]). They also studied the effect of different counterions on the polarizability. No differences were noted for the monovalent cations Li^+, Na^+, and K^+, although the polarizability was lower for the tetramethylammonium cation. The polarizability was much higher with Mg^{2+} as the counterion, but decreased rapidly with increasing Mg^{2+} ion concentration.

Evidence about the orienting mechanism of DNA molecules can also be obtained from electric birefringence experiments in which the polarity of the applied field is reversed after steady state orientation is achieved (see Chap. 3). Hanss et al. [110] reported that a negative transient was observed in DNA solutions at the moment of field reversal. Stellwagen [37] found that negative or positive transients could be observed, depending on the buffer concentration and electric field strength used. The negative transient observed in solutions of high buffer concentration can probably be attributed to a slow relaxation of the ion atmosphere along the long axis of the DNA molecules; the positive transient which was observed in solutions of low buffer concentration may indicate a contribution from a permanent dipole moment in the transverse direction [37].

G.   Chemical Kinetic Measurements

Several kinetic studies have used electric birefringence as an analytical tool. Norman and Field [111] studied the irradiation of DNA with x-rays and found that the larger molecules were eliminated preferentially. The decay of the birefringence of the irradiated particles could be described by a single relaxation time. Ingram and Jerrard [56] studied the action of deoxyribonuclease on DNA in the presence of Mg^{2+}. After an induction period the birefringence decay times gradually decreased and then leveled off; there was very little change in the magnitude of the birefringence with time. Itzhaki [66] studied the structure of deoxyribonucleoprotein and DNA by analyzing the changes in the birefringence produced by deoxyribonuclease and proteolytic enzymes. She concluded that DNA molecules in deoxyribonucleoprotein were not linked lengthwise by protein molecules.

III.   RIBONUCLEIC ACID

The electric birefringence of RNA in solution was first studied by Ginoza and Norman [112], who observed that free RNA from tobacco mosaic virus exhibited positive birefringence. Since positive birefringence had also been observed in the virus [113], they inferred that the purine and pyrimidine bases in free RNA were oriented parallel to the long axis of the molecules. The relaxation time indicated a length close to that of the virus itself.

Ingram and Jerrard [55] studied yeast RNA in 50% water/ glycerol solutions and observed that the decay of the birefringence could be described by two relaxation times. They also noted that the kinetics of the degradation of RNA by ribonuclease could easily be followed by Kerr effect measurements.

Morgan [98] used electric birefringence and ultraviolet electric dichroism to study the orientation of RNA within 80S yeast ribosomes. Positive birefringence and parallel dichroism were observed, indicating that the purine and pyrimidine bases were oriented with transition moments mainly parallel to the axis of electrical anisotropy of the ribosomes. Morgan proposed that, within the ribosomes, intramolecular helical segments of RNA were oriented perpendicular to the long axis of the ribosomes. A similar structure has been proposed by Spirin [18] for free RNA molecules in solution (see Fig. 6b).

Golub and Dvorkin [114] studied the electric birefringence and dichroism of tobacco mosaic virus RNA as a function of wavelength. They concluded that dichroism was important only near the strong ultraviolet absorption band and that the contribution of dichroism to the birefringence signal observed in the visible region of the spectrum was negligible.

Golub and Nazarenko [50,115] studied the electric birefringence of high molecular weight RNAs obtained from tobacco mosaic virus, E. coli, ascites tumor, rat liver, and calf thymus. Stellwagen [37], using electric birefringence, and Frisman et al. [116], using flow birefringence, also studied ascites tumor RNA. Golub and Nazarenko [50,115] and Stellwagen [37] both observed the simultaneous presence of positive and negative components in the birefringence signals of RNA solutions. The relaxation time of the positive component was of the order of hundreds of microseconds [37,50]; the relaxation time of the negative component was much smaller, typically about 10 μsec [37].

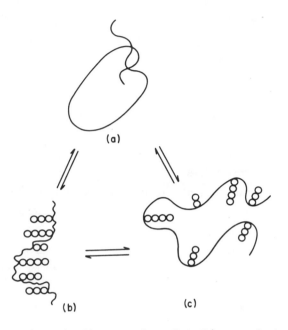

FIG. 6.   Schematic diagram of conformations and structural transitions of high molecular weight RNA molecules in solution [after A. S. Spirin, J. Mol. Biol., 2, 436 (1960)].   (a) extended coil, positive birefringence; (b) compact rod, positive birefringence; (c) compact coil, negative birefringence.

Golub and Nazarenko [50] found that E. coli RNA exhibited negative birefringence at room temperature, while tobacco mosaic virus, ascites tumor, rat liver, and calf thymus RNA exhibited both positive and negative birefringence.   The relative magnitude of the positive and negative components depended on the temperature, ionic strength, and concentration of divalent cations. Increasing the temperature or removing Mg^{2+} enhanced the positive birefringence of the solutions; at sufficiently high temperatures only positive birefringence was observed.   Decreasing the temperature or increasing the Mg^{2+} concentration enhanced the negative birefringence of the solutions.   Stellwagen [37] found that the relative magnitude of the positive and negative components in the birefringence signal depended on the RNA concentration.   Negative birefringence was observed in very dilute solutions of ascites tumor RNA (0.01 g/liter in 10^{-4} M Tris buffer), but solutions of higher concentration (0.04 g/liter or greater in 4×10^{-4} M Tris buffer) exhibited both positive and negative birefringence. Frisman et al. [116] found that ascites tumor RNA exhibited positive flow birefringence at low shear gradients and negative birefringence at high shear gradients.

Electric dichroism of tobacco mosaic virus RNA was studied by Dvorkin and Spirin [18,19]. At low temperatures and in solutions of low ionic strength (0.01 M), positive dichroism was observed with a dichroic ratio of +1.2 (see Chap. 7), indicating that the purine and pyrimidine bases were oriented mainly parallel to the long axis of the RNA molecules. The dichroic ratio remained constant from 10-30°C, disappeared at 40°C, and became negative at 50°C, indicating that at this temperature the purine and pyrimidine bases were oriented mainly perpendicular to the long axis of the molecules. This change in sign of the dichroism was reversible. When the ionic strength was increased to 0.1, the dichroic ratio observed at 20°C decreased to +1.1, indicating a less ordered arrangement of the bases at this ionic strength.

The birefringence and dichroism results described above, as well as other hydrodynamic properties [18], can be reconciled by the conformations and structural transitions proposed by Spirin [18] and illustrated schematically in Fig. 6. At room temperature in solutions of low ionic strength the conformation of RNA is proposed to be that of a compact rod, structure (b) in Fig. 6, with helical segments oriented perpendicular to the rod axis. In this type of structure the purine and pyrimidine bases are oriented parallel to the rod axis, giving rise to the positive dichroism observed by Dvorkin and Spirin [18,19] and the positive birefringence observed by Stellwagen [37], Morgan [98], and Frisman et al. [116]. The helical segments may be held together by chelation with Mg^{2+}; Haschemeyer et al. [117] reported that two RNA conformations can exist in aqueous solution depending on the presence

or absence of Mg^{2+} (transition $b \rightleftharpoons a$ in Fig. 6). When the RNA
concentration is decreased, as in the birefringence experiment
[37], or the temperature increased, as in the dichroism experiment
[19], the tertiary structure is disrupted and the conformation of
the RNA molecules changes to that of the compact coil, structure
c (transition $b \rightleftharpoons c$ in Fig. 6). In this conformation the
secondary structure still exists, but the helical segments are
randomly oriented. Since the helical segments have a higher
charge density than random coil sections of the molecule, the
helical segments would orient preferentially in an electric field,
leading to the perpendicular dichroism observed by Dvorkin and
Spirin [18,19] at $50°C$ and the negative birefringence observed at
room temperature by Golub and Nazarenko [50] and Stellwagen [37].
The very fast rise and decay times (≤ 10 μsec) of the negative
birefringence signals are consistent with the orientation of
individual helical segments in the electric field. The negative
birefringence observed by Golub and Nazarenko [50] for some RNA
preparations could be changed to positive birefringence by decreasing the Mg^{2+} concentration or by heating the solutions above
the melting temperature (transition $c \rightleftharpoons a$ in Fig. 6). The positive birefringence observed by Golub and Nazarenko [50] at high
temperatures was undoubtedly due to the extended coil form of the
molecule, structure a in Fig. 6. Since a relaxation time of
several hundred microseconds was observed, the molecule must have
been fairly extended, probably because of mutual repulsions of the
negative charges on the backbone chain. The negative flow birefringence observed at high shear gradients by Frisman et al. [116]
was probably due to orientation of helical RNA fragments cleaved
by the high shear gradients.

The positive and negative components of the birefringence of
various RNA samples seem to exhibit different types of saturation
behavior. Morgan [98] found that the positive birefringence
exhibited by tobacco mosaic virus RNA in ribosomes increased
linearly with electric field strength. Golub and Nazarenko [50]

found that the negative components of the birefringence of rat
liver, calf thymus, and tobacco mosaic virus RNA in water were
proportional to the square of the electric field strength, while
the positive birefringence exhibited by a preparation of tobacco
mosaic virus precipitated from 0.01 M EDTA and redissolved in
water was proportional to $E^{1.1}$. Stellwagen [37] found that the
negative birefringence observed for ascites tumor RNA in 10^{-4} M
Tris buffer was proportional to $E^{1.3}$.

The apparent Kerr constants calculated from the negative
birefringence of ribonucleic acids obtained from different sources
are summarized in Table 6.

Low molecular weight ribonucleic acids (transfer RNA; soluble
RNA in older terminology) have been studied only by flow birefringence. Transfer RNAs obtained from yeast, E. coli, and rabbit
liver all exhibit negative birefringence [25,118], indicating that
the purine and pyrimidine bases are oriented mainly perpendicular
to the long axis of the RNA molecules.

TABLE 6

Apparent Kerr Constants[a] Calculated from
Negative Birefringence of Different RNA Samples

Investigator	Ref.	RNA source	Solvent	$-K_{app}$[a] $\times 10^6$ (cgs esu)
Golub and Nazarenko	50,115	E. coli	Water	35
		Rat liver	Water	4.7
		Calf thymus	Water	1.6
Stellwagen	37	Ascites tumor	10^{-4} M Tris	8.7

[a]Calculated as though the Kerr law were obeyed.

IV.   SYNTHETIC POLYNUCLEOTIDES

The electric birefringence of polyadenylic acid (poly A), polyuridylic acid (poly U), and the complexes formed by the inter- action of these two polynucleotides, was studied by Jakabhazy and Fleming [119] and by Stellwagen [37]. Stellwagen [37] also studied polyinosinic acid (poly I), polycytidylic acid (poly C), and the poly (I + C) complex. Typical data for the birefringence of these synthetic polynucleotides and complexes are presented in Table 7.

Poly A, poly C, and the poly (A + U), poly (A + 2U), and poly (I + C) complexes all exhibited negative birefringence, indicating that the purine and pyrimidine bases were oriented mainly perpendi- cular to the long axis of the molecules. Negative flow birefrin- gence has also been observed for poly A [120] and the poly (A + U) [120] and poly (I + C) [120a] complexes. The negative birefrin- gence observed for poly A and poly C is consistent with other evidence indicating that the conformation of these molecules is that of a single-stranded helix stabilized by base stacking [28, 29,121,122]. Poly U and poly I exhibited positive birefringence (Table 7), indicating that the bases were either randomly oriented or oriented mainly parallel to the long axis of the molecules. The positive birefringence observed for poly U at 4°C indicates that the low temperature form of this molecule is not a double helix, as originally proposed [123]; such a structure would exhibit negative birefringence. A structure similar to that proposed for high molecular weight RNA (structure b in Fig. 6), with helical segments oriented perpendicular to the long axis of the molecules, would be consistent with the positive birefringence observed at 4°C [119].

Jakabhazy and Fleming [119] utilized the large increase in the magnitude of the birefringence of the poly (A + U) complex (see Table 7) to study the rate of formation of the complex. In aqueous solutions containing an equivalent concentration of Mg^{2+},

TABLE 7

Electric Birefringence of Synthetic Polynucleotides and Polynucleotide Complexes

Investigator	Ref.	Polynucleotide	Conc. (g/liter)	Solvent	Temp. (°C)	$K_{app}^a \times 10^6$ (cgs esu)	t_1^b (μsec)
Jakobhazy and Fleming	119	Poly A	0.13	Water + Mg^{2+}	4	-70	360
		Poly U	0.13	Water + Mg^{2+}	4	+3.2	~250
		Poly (A + U)	0.13	Water + Mg^{2+}	4	-580	700
		Poly (A + 2U)	0.13	Water + Mg^{2+}	4	-1100	800
Stellwagen	37	Poly I	0.04	4×10^{-4} M Tris	25	+1.5	6
		Poly C	0.04	4×10^{-4} M Tris	25	-1.7	12
		Poly (I + C)	0.02	4×10^{-4} M Tris	25	-22	20

a Apparent Kerr constant, since the Kerr law may not be obeyed for these systems [37].

b Decay time, defined as the time required for the birefringence to decrease to 1/e of its steady state value.

formation of the complex was found to follow second order kinetics
up to 60-70% conversion of the poly A or poly U to poly (A + U).
A second order reaction is consistent with a reaction mechanism
involving a bimolecular "nucleation" step followed by rapid
propagation of base pairs along the backbone chain. From an analysis of the data based on the Brønsted-Bjerrum theory [124,125],
Jakabhazy and Fleming estimated that about 7 A-U base pairs were
involved in the activated complex.

Jakabhazy and Fleming [119] also studied the dependence of
the specific Kerr constant and decay times of poly A on pH, using
a 25 kc/sec alternating electric field. No birefringence was observed in acid solution. As the pH was increased, the birefringence increased sharply between pH 5 and 6.5, with a midpoint at
pH 6, indicating that a highly cooperative transition had
occurred. The decay times, and hence the length of the poly A
molecules, remained essentially constant throughout this pH range.
Therefore, the large increase in the specific Kerr constant must
have been due to the increased polarization of the ion atmosphere,
which paralleled the increased net negative charge on the molecules. Any permanent dipole moment would not have contributed to
the birefringence in the rapidly alternating electric field (see
Chap. 3). Hence this increase in K_{sp} with increasing pH fur-
nishes a proof of the ion atmosphere polarization mechanism.

V.   CONCLUDING REMARKS

From the results presented in this chapter it is apparent
that electro-optic methods provide a powerful tool for studying
conformational changes and investigating the electric and optic
properties of macromolecules in solution. One of the important
problems remaining is the investigation of the non-Kerr law
behavior of DNA and other polyelectrolytes. Although this effect

has been noted by several investigators studying a variety of
polyelectrolytes, no theoretical explanation is yet available.

Electro-optic techniques seem ready to yield useful information on many other problems. The conformational changes of high
molecular weight RNA should be studied systematically as a function of temperature, RNA concentration, and Mg^{2+} concentration.
Soluble 5S ribosomal RNA might provide an interesting system to
study; it is thought to have a structure of relatively low axial
ratio (a/b = 5:1) with helical segments oriented parallel to the
long axis of the molecule [126]. The thermal denaturation of
transfer RNA seems to be a two step process, involving a conformational change at temperatures below which changes in optical
density occur [127,128].

Conformational changes in DNA are not completely understood.
At temperatures well below the melting temperature, where little
or no change in optical density occurs, large changes in circular
dichroism [129] and in the magnitude of the flow dichroism [52]
and flow birefringence [42] have been observed. These effects
have been interpreted in terms of an intermediate conformation of
DNA [129-132], but a partial unwinding of the DNA helix [88] might
also be considered. It should be possible to observe the kinetics
of the denaturation and renaturation of DNA by electro-optic
techniques. The rate of denaturation is inversely proportional to
molecular weight, but the rate of renaturation is independent of
molecular weight, suggesting that an intramolecular activation
step may precede the rewinding of the DNA molecules [133].
Conformational changes in DNA have been noted in the presence of
Cu^{2+} [134-136] and other heavy metal ions [136,137], in ethanol
[49,53,138], and in ethylene glycol [139]. A "condensed" form of
DNA has been reported at low pH [140]. DNA forms stoichiometric
complexes with certain proteins and polycations such as chymotrypsin [141,142], spermine [143,144], polylysine [145-148,148a],
polyarginine [148], and polyvinylamine [145]. Polylysine also
forms stoichiometric complexes with poly (A + U) [148a,149] and
poly (I + C) [148a]. The oligopeptide antibiotic netropsin binds

to DNA causing elongation and stiffening of the DNA without intercalation of the netropsin [150]. The kinetics of formation of the poly (I + C) complex could also be studied. The heat of reaction between poly I and poly C is independent of salt concentration and temperature, unlike the heat of reaction between poly A and poly U [151].

ACKNOWLEDGMENT

Helpful comments from L. L. Mack are gratefully acknowledged.

REFERENCES

1. J. D. Watson and F. H. C. Crick, Nature, 171, 737 (1953).

2. R. Langridge, H. R. Wilson, C. W. Hooper, M. H. F. Wilkins, and L. D. Hamilton, J. Mol. Biol., 2, 19 (1960).

3. R. Langridge, D. A. Marvin, W. E. Seeds, H. R. Wilson, C. W. Hooper, M. H. F. Wilkins, and L. D. Hamilton, J. Mol. Biol., 2, 38 (1960).

4. S. Bram and W. W. Beeman, J. Mol. Biol., 55, 311 (1970).

5. V. Luzzati, A. Nicolaieff, and F. Masson, J. Mol. Biol., 3, 185 (1961).

6. C. L. Sadron, in The Nucleic Acids (E. Chargaff and J. N. Davidson, eds.), Vol. III, Academic Press, New York, 1960, pp. 15-37.

7. J. Josse and J. Eigner, Ann. Rev. Biochem., 35, part II, pp. 789-834 (1966).

8. V. A. Bloomfield, Macromol. Rev., 3, 255 (1968).

9. D. M. Crothers and B. H. Zimm, J. Mol. Biol., 12, 525 (1965).

10. E. V. Frisman, V. I. Vorob'ev, and L. V. Shchagina, Vysokomolekul. Soedin, 6, 884 (1964).

11. G. Weill, C. Hornick, and S. Stoylov, J. Chim. Phys., 64, 182 (1968).

12. O. Kratky and G. Porod, Rec. Trav. Chim. Pays-Bas, 68, 1106 (1949).

13. B. H. Zimm, J. Chem. Phys., 24, 269 (1956).

14.   P. E. Rouse, J. Chem. Phys., 21, 1272 (1953).

15.   F. Bueche, J. Chem. Phys., 22, 603 (1954).

16.   P. Doty, H. Boedtker, J. R. Fresco, R. Haselkorn, and
      M. Litt, Proc. Nat. Acad. Sci. Wash., 45, 482 (1959).

17.   J. R. Fresco, B. M. Alberts, and P. Doty, Nature, 188, 98
      (1960).

18.   A. S. Spirin, J. Mol. Biol., 2, 436 (1960).

19.   G. A. Dvorkin and A. S. Spirin, Dokl. Akad. Nauk SSSR, 135,
      987 (1960).

20.   E. S. Bogdanova, L. P. Gavrilova, G. A. Dvorkin, N. A.
      Kisselev, and A. S. Spirin, Biokhimiya, 27, 387 (1962).

21.   A. S. Spirin, in Progress in Nucleic Acid Research (J. N.
      Davidson and W. E. Cohn, eds.), Vol. 1, Academic Press, New
      York, 1963, pp. 301-341.

22.   N. A. Kisselev, L. P. Gavrilova, and A. S. Spirin, J. Mol.
      Biol., 3, 778 (1961).

23.   S. N. Timasheff, J. Witz, and V. Luzzati, Biophys. J., 1, 525
      (1961).

24.   S. N. Timasheff, Biochim. Biophys. Acta, 88, 630 (1964).

25.   V. N. Tsvetkov, L. L. Kiselev, L. Y. Frolova, and S. Y.
      Lyubina, Vysokomolekul. Soedin, 6, 568 (1964).

26.   M. Levitt, Nature, 224, 759 (1969).

27.   J. D. Watson, Molecular Biology of the Gene, 2nd ed., W. A.
      Benjamin, New York, 1970, pp. 346-7; 358-364.

28.   V. Luzzati, J. Witz, and A. Mathis, in Genetic Elements
      (D. Shugar, ed.), Academic Press, London, 1967, pp. 41-55.

29.   J. T. Yang and T. Samejima, Progr. Nucleic Acid Research and
      Mol. Biol., 9, 223 (1969).

30.   H. Benoit, J. Chim. Phys., 48, 612 (1951).

31.   C. T. O'Konski and A. J. Haltner, Abstr. of Amer. Chem. Soc.,
      125th Meeting, Kansas City, 1954, p. 9R.

32.   A. J. Haltner, Ph.D. Thesis, Univ. of Calif., Berkeley, 1955.

33.   G. A. Dvorkin and E. I. Golub, Biofizika, 8, 301 (1963).

34.   E. I. Golub, Biopolym., 2, 113 (1964).

35.   C. Houssier and E. Fredericq, Biochim. Biophys. Acta, 88, 450
      (1964).

36.   N. C. Stellwagen, M. Shirai, and C. T. O'Konski, Abstracts,
      9th Biophysical Society Meeting, San Francisco, 1965.
      (Abstract WG-10).

37. N. C. Stellwagen, Ph.D. Thesis, Univ. of Calif., Berkeley, 1967.

38. J. C. Powers, Jr. and W. L. Peticolas, J. Phys. Chem., 71, 3191 (1967).

39. C. Houssier and H.-G. Kuball, Biopolym., 10, 2421 (1971).

40. R. Signer, T. Caspersson, and E. Hammersten, Nature, 141, 122 (1938).

41. O. Snellman and G. Widström, Arkiv. Kemi Mineral. Geol., 19A, No. 31 (1945).

42. E. V. Frisman, V. I. Vorob'yev, L. V. Shchagina, and N. K. Yanovskaya, Vysokomolekul. Soedin, 5, 622 (1963).

43. R. E. Harrington, Biopolym., 9, 159 (1970).

43a. G. B. Thurston and R. S. Wilkinson, Biorheology, 10, 351 (1973).

44. W. T. Astbury and F. O. Bell, Nature, 141, 747 (1938).

45. W. E. Seeds and M. H. F. Wilkins, Disc. Faraday Soc., No. 9, 417 (1950).

46. G. A. Dvorkin, Biofizika 6, 403 (1961).

47. G. A. Dvorkin, Dokl. Akad. Nauk SSSR, 135, 739 (1960).

48. G. A. Dvorkin and V. I. Krinskii, Dokl. Akad. Nauk SSSR, 140, 942 (1961).

49. D.-W. Ding, R. Rill, and K. E. Van Holde, Biopolym., 11, 2109 (1972).

50. E. I. Golub and V. G. Nazarenko, Biophys. J., 7, 13 (1967).

51. C. T. O'Konski, N. C. Stellwagen, and M. Shirai, Biophys. J., accepted for publication.

52. L. F. Cavalieri, M. Rosoff, and B. H. Rosenberg, J. Am. Chem. Soc., 78, 5239 (1956).

53. J. Brahams and V. F. H. M. Mommaerts, J. Mol. Biol., 10, 73 (1964).

54. C. T. O'Konski and A. J. Haltner, J. Am. Chem. Soc., 78, 3604 (1956).

55. P. Ingram and H. G. Jerrard, Nature, 196, 57 (1962).

56. P. Ingram and H. G. Jerrard, Brit. J. Appl. Phys., 14, 572 (1963).

57. B. R. Jennings and H. Plummer, Biopolym., 9, 1361 (1970).

58. E. I. Golub, G. A. Dvorkin, and V. G. Nazarenko, Biokhimiya, 28, 1041 (1963).

59. M. L. Allais, J. Chim. Phys., 59, 873 (1962).

60. C. Hornick and G. Weill, Biopolym., 10, 2345 (1971).

61. L. F. Cavalieri, B. H. Rosenberg, and M. Rosoff, J. Am. Chem. Soc., 78, 5235 (1956).

62. H. Schwander and R. Cerf, Helv. Chim. Acta, 32, 2356 (1949).

63. A. R. Mathieson and S. J. Matty, J. Polym. Sci., 23, 747 (1957).

64. N. Ise, M. Eigen, and G. Schwarz, Biopolym., 1, 343 (1963).

65. M. Goldstein and M. E. Reichman, J. Am. Chem. Soc., 76, 3337 (1954).

66. R. F. Itzhaki, Proc. Roy. Soc. B, 164, 411 (1966).

67. H. H. E. W. Ohlenbusch, Ph.D. Thesis, Cal. Tech., Pasadena, 1966; Diss. Abstr., 27, 725 (1966).

68. R. E. Harrington, Biopolym., 6, 105 (1968).

69. P. R. Callis and N. Davidson, Biopolym., 8, 379 (1969).

70. D. S. Thompson and S. J. Gill, J. Chem. Phys., 47, 5008 (1967).

71. R. E. Chapman, Jr., L. C. Klotz, D. S. Thompson, and B. H. Zimm, Macromol., 2, 637 (1969).

72. H. R. Massie and B. H. Zimm, Biopolym., 7, 475 (1969).

73. K. E. Reinert, J. Strassburger, and H. Triebel, Biopolym., 10, 285 (1971).

74. V. N. Tsvetkov, L. N. Andreyeva, and L. N. Kvitchenko, Vysokomolekul. Soedin, 7, 2001 (1965).

75. V. N. Tsvetkov, in Newer Methods of Polymer Characterization, (B. Ke, ed.), Interscience, New York, 1963, Chap. 14.

76. V. N. Tsvetkov, Vysokomolekul. Soedin, 5, 740,747 (1963).

77. V. N. Tsvetkov, Vysokomolekul. Soedin, 7, 1468 (1965).

78. V. N. Tsvetkov, L. N. Andreyeva, and V. I. Sisenko, Molekul. Biofiz. Akad. Nauk SSSR, Inst. Biol. Fiz. Static, 1965, pp. 110-116.

79. J. A. Harpst, A. I. Krasna, and B. H. Zimm, Biopolym., 6, 585,595 (1968).

80. S. Broersma, J. Chem. Phys., 32, 1626 (1960).

81. H. B. Gray, V. A. Bloomfield, and J. E. Hearst, J. Chem. Phys., 46, 1493 (1967).

82. J. E. Hearst, C. W. Schmid, and F. P. Rinehart, Macromol., 1, 491 (1968).

83. J. B. Hays, M. E. Magar, and B. H. Zimm, Biopolym., 8, 531 (1969).

84. A. Sharp and V. A. Bloomfield, Biopolym. 6, 1201 (1968).

85. C. W. Schmid, F. P. Rinehart, and J. E. Hearst, Biopolym., 10, 883 (1971).

86. H. Triebel and K. E. Reinert, Biopolym., 10, 827 (1971).

87. D. Lang, A. K. Kleinschmidt, and R. K. Zahn, Biochim. Biophys. Acta, 88, 142 (1964).

88. D. Lang, H. Bujard, B. Wolff, and D. Russell, J. Mol. Biol., 23, 163 (1967).

89. I. Isenberg and R. Dyson, Biophys. J., 9, 1337 (1969).

90. C. T. O'Konski, K. Yoshioka, and W. H. Orttung, J. Phys. Chem., 63, 1558 (1959).

91. C. T. O'Konski and N. C. Stellwagen, Biophys. J., 5, 607 (1965).

91a. K. Yamaoka and E. Charney, Macromol., 6, 66 (1973).

92. E. Neumann and A. Katchalsky, Proc. Nat. Acad. Sci. Wash., 69, 993 (1972).

93. J. M. Burgers, "Verhandel. Koninkl. Ned. Akad. Wetenschap. Afdeel. Natuurk.," Sec. 1, Dell XVI, No. 4, 113 (1938).

94. J. Kerr, Phil. Mag., 50, 337,446 (1875).

95. H. Benoit, Ann. Phys., 6, 561 (1951).

96. H. Benoit, J. Chim. Phys., 47, 719 (1950).

97. T. Soda and K. Yoshioka, Nippon Kagaku Zasshi, 87, 1326 (1966).

98. R. S. Morgan, Biophys. J., 3, 253 (1963).

99. C. Houssier and E. Fredericq, Biochim. Biophys. Acta, 120, 113 (1966).

100. H. Nakayama and K. Yoshioka, Nippon Kagaku Zasshi, 85, 177, 258 (1964); J. Polym. Sci., Part A, 3, 813 (1965).

101. K. Kikuchi and K. Yoshioka, Reports on Progress in Polymer Physics in Japan, 10, 19 (1967); further results to be published.

102. C. T. O'Konski and B. H. Zimm, Science, 111, 113 (1950).

103. C. T. O'Konski and A. J. Haltner, J. Am. Chem. Soc., 79, 5634 (1957).

104. C. T. O'Konski, J. Phys. Chem., 64, 605 (1960).

105. C. T. O'Konski and S. Krause, J. Phys. Chem., 74, 3243 (1970).

106. H. A. Stuart and A. Peterlin, J. Polym. Sci., 5, 551 (1950).

107. A. Scheludko and S. Stoylov, Biopolym., 5, 723 (1967).

108. S. P. Stoylov, Adv. Colloid Interface Sci., $\underline{3}$, 45 (1971).

109. G. Weill and C. Hornick, Biopolym., $\underline{10}$, 2029 (1971).

110. M. Hanss, J. C. Bernengo, and B. Roux, results presented at
      XXIII International Congress of Pure and Applied Chemistry,
      Boston (1971), cited by reference 49.

111. A. Norman and J. A. Field, Jr., Arch. Biochem. Biophys., $\underline{71}$,
      170 (1957).

112. W. Ginoza and A. Norman, Nature, $\underline{179}$, 520 (1957).

113. R. E. Franklin, Biochim. Biophys. Acta, $\underline{18}$, 313 (1955).

114. E. I. Golub and G. A. Dvorkin, Biofizika, $\underline{9}$, 545 (1964).

115. E. I. Golub and V. G. Nazarenko, Biofizika, $\underline{9}$, 657 (1964).

116. E. V. Frisman, V. I. Vorob'yev, N. K. Yanovskaya, and
      L. V. Shchagina, Biokhimiya, $\underline{28}$, 137 (1963).

117. R. Haschemeyer, B. Singer, and H. Fraenkel-Conrat, Proc. Nat.
      Acad. Sci. Wash., $\underline{45}$, 313 (1959).

118. V. N. Tsvetkov, L. L. Kiselev, S. Y. Lyubina, L. Y. Frolova,
      S. I. Klenin, V. S. Skozka, and N. A. Nikiten, Biokhimiya,
      $\underline{30}$, 302 (1965).

119. S. Z. Jakabhazy and S. W. Fleming, Biopolym., $\underline{4}$, 793 (1966).

120. S. Takashima, Biopolym., $\underline{6}$, 1437 (1968).

120a. H. Akutsu, M. Fuke, and H. Hashizume, J. Biochem. (Tokyo),
      $\underline{74}$, 827 (1973).

121. A. Gulik, H. Inoue, and V. Luzzati, J. Mol. Biol., $\underline{53}$, 221
      (1970).

122. K. G. Brown, E. J. Kiser, and W. L. Peticolas, Biopolym., $\underline{11}$,
      1855 (1972).

123. M. N. Lipsett, Proc. Nat. Acad. Sci. Wash., $\underline{46}$, 445 (1960).

124. J. N. Brønsted, Z. physik. Chem., $\underline{102}$, 169 (1922); $\underline{115}$, 337
      (1925).

125. N. Bjerrum, Z. physik. Chem., $\underline{108}$, 82 (1924).

126. P. G. Connors and W. W. Beeman, J. Mol. Biol., $\underline{71}$, 31 (1972).

127. D. D. Henley, T. Lindahl, and J. R. Fresco, Proc. Nat. Acad.
      Sci. Wash., $\underline{55}$, 191 (1966).

128. D. B. Millar and R. F. Steiner, Biochem., $\underline{5}$, 2289 (1966).

129. D. S. Studdert, M. Patroni, and R. C. Davis, Biopolym., $\underline{11}$,
      761 (1972).

130. V. Luzzati, A. Mathis, F. Masson, and J. Witz, J. Mol. Biol.,
      $\underline{10}$, 28 (1964).

131. J. Marmur, R. Rownd, and C. L. Schildkraut, in Progress in Nucleic Acid Research (J. N. Davidson and W. E. Cohn, eds.), Vol. 1, Academic Press, New York, 1963, pp. 231-293.

132. P. Bartl and M. Boublík, Biochim. Biophys. Acta, 103, 678 (1965).

133. M. T. Record and B. H. Zimm, Biopolym., 11, 1435 (1972).

134. C. Zimmer, G. Luck, H. Fritzsche, and H. Triebel, Biopolym., 10, 441 (1971).

135. D. C. Liebe and J. E. Stuehe, Biopolym., 11, 145,167 (1972).

136. J. P. Schreiber and M. Daune, Biopolym., 8, 139 (1969).

137. J. A. Anderson, G. P. P. Kuntz, H. H. Evans, and T. J. Swift, Biochem., 10, 4368 (1971).

138. D. Lang, J. Mol. Biol., 46, 209 (1969).

139. G. Green and H. R. Mahler, Biopolym., 6, 1509 (1968).

140. E. Dore, C. Frontali, and E. Gratton, Biopolym., 11, 443 (1972).

141. B. H. J. Hofstee, Biochim. Biophys. Acta, 55, 440 (1962).

142. B. H. J. Hofstee, J. Biol. Chem., 238, 3235 (1963).

143. H. Tabor, Biochem., 1, 496 (1962).

144. M. Suwalsky, W. Traub, U. Shmueli, and J. A. Subirana, J. Mol. Biol., 42, 363 (1969).

145. P. Spitnik, R. Lipshitz, and E. Chargaff, J. Biol. Chem., 215, 765 (1955).

146. D. E. Olins, A. L. Olins, and P. H. von Hippel, J. Mol. Biol., 24, 157 (1967).

147. J. L. Shaprio, M. Leng, and G. Felsenfeld, Biochem., 8, 3219 (1969).

148. M. Suwalsky and W. Traub, Biopolym., 11, 2223 (1972).

148a. M. Haynes, R. A. Garrett, and W. B. Gratzer, Biochem., 9, 4410 (1970).

149. G. Felsenfeld and S. Huang, Biochim. Biophys. Acta, 34, 234 (1959).

150. K. E. Reinert, J. Mol. Biol., 72, 593 (1972).

151. P. D. Ross and R. L. Scruggs, J. Mol. Biol., 45, 567 (1969).

Chapter 19

ELECTRO-OPTICS OF
POLYELECTROLYTES AND DYE-POLYELECTROLYTE COMPLEXES

Michio Shirai

Department of Chemistry
College of General Education
University of Tokyo
Komaba, Meguro, Tokyo

I. INTRODUCTION

Many macromolecules exist in aqueous solution in the form of macroions. Proteins, nucleic acids, and many synthetic polymers are included in this group, and they are called polyelectrolytes. The solutions of polyelectrolyte must be electrically neutral, so that each macroion contains a number of small ions of opposite charge in its vicinity to neutralize the charge of the macroion. These small counterions play an important role in the electro-optic properties of polyelectrolyte solutions. There is strong electrostatic interaction between charges on macroions and small ions, and the small counterions form the diffuse double layer around a macroion.

Early studies on colloidal solutions showed that V_2O_5 sols display a positive birefringence [1]. In a later study on the electric birefringence of V_2O_5 sol in sine wave fields [2] it was found that the Kerr effect was divided into a positive component due to double-layer polarization and a negative part arising from a permanent moment perpendicular to the long axis.

The electric birefringence of tobacco mosaic virus (TMV) solutions was measured to elucidate the nature of the orienting forces in an electric field [3-5]. The effect of ionic strength on the electric birefringence, the reversing field behavior, the dispersion, the buildup behavior, all suggested that the forces arise from ion atmosphere polarization, though it had previously been thought to arise from the permanent dipole moment [6]. The field strength dependence of the birefringence also showed that it is due to a large ion atmosphere polarizability, and permanent dipole effect is not significant [5]. For polyethylene sulfonate, however, the permanent dipole effect was found to be dominant, although it is a polyelectrolyte and must have a large ionic polarizability [5,7].

The electro-optic properties of many other polyelectrolytes have been investigated. The mechanism of orientation in the electric field has been a most interesting problem. For most polyelectrolytes the orientation is due to counterion polarization, which is due to the ionic transport arising from small ions on and near the surfaces of macroions. This phenomenon was treated by O'Konski in connection with dielectric and conductivity properties [8]. Variations of this theory were proposed to explain the large electric polarization of polyelectrolyte solutions [9,10]. A theory applicable to the electro-optics was developed by O'Konski and Krause [11]. Effects of ionic strength, pH, and field strength on the electric birefringence were explained by these theories.

The electro-optic properties of complexes of polyelectrolytes with dyes were studied, and the orientation of chromophores in the complexes were determined [12,13]. These results also supplied useful knowledge concerning the structure of complexes or polyelectrolytes.

II. ELECTRO-OPTIC BEHAVIOR OF POLYELECTROLYTES

In interpretations of the electro-optic behavior of polyelectrolytes the mechanism of electric polarization is an important problem. The orientation in an electric field is due to ion atmosphere polarization or permanent dipole moment, but proton polarization, discussed by Kirkwood and Shumaker [14], is also possible. There are several experimental methods of determining which mechanism is important for a polyelectrolyte. We will review the methods of determining the mechanism of orientation and the results for various polyelectrolytes.

MICHIO SHIRAI

A. Buildup and Decay

O'Konski and Zimm [3] first reported that the birefringence
relaxation time τ is given by $\tau = 1/6\theta$. Near the same time,
Benoit [15], obtained equations for both the buildup and decay
curves of birefringence with a square pulse of a weak field.
According to his theory, the buildup and decay curves are symmetrical for induced polarization, whereas the buildup process is
slower than the decay process for permanent dipole orientation.
Tinoco [16], extended the theory to the more general case where
the direction of the permanent moment is not along the hydrodynamic symmetry axis. O'Konski et al. [5], derived the buildup equations of birefringence for permanent dipole and induced dipole
orientation in strong electric fields. They showed that the
initial slope of the buildup curve for permanent dipole orientation is zero. Birefringence buildup of TMV is of the induced
dipole orientation type, and results in the same time constant as
that for the decay 0.50 msec [4].

B. Birefringence in a Reversing Field

It is interesting to detect the transient in the
birefringence which occurs upon reversal of the polarizing field.
Upon reversal of the field, TMV solutions showed no transient [4].
This result leads to the conclusion that orientation of TMV is
due to an induced polarization, which is reversed by reversing the
applied field, and the orienting torque remains unchanged upon
field reversal. If orientation is a result of permanent dipole
moment the birefringence signal will drop upon field reversal.

Tinoco and Yamaoka [17], derived the equation for the birefringence in a rapidly reversing field, and the contributions of
permanent and induced polarization are determined quantitatively.
The difference between the signals for the two mechanisms is

significant, so that this method is more effective than the
analysis of the buildup curve to determine the mechanism of
orientation.

A similar case of interest is that of a pulse consisting of
several oscillations of a sine wave [4,18]. At sufficiently high
frequencies of a sine wave a steady birefringence arises from an
induced dipole and the birefringence due to permanent dipole
disappears (see Chap. 3). Therefore, with this waveform only the
birefringence arising from induced dipole can be measured.

C. Saturation Effect

In weak electric fields the Kerr constant is proportional to
the square of the field strength. But the saturation effect of
the orientation occurs in strong fields. The exact calculation
of field dependence in arbitrary field strength was treated by
O'Konski et al. [5]. An experimental plot of field strength of
the birefringence may be made to fit a theoretical curve by
adjusting the parameters. From calculated values of the parameters the contributions of permanent dipole and induced dipole are
determined.

O'Konski et al. [5], studied the orienting mechanism of some
polyelectrolytes by making use of the saturation effect. They
found that TMV exhibited saturation characteristic of a large
polarizability, and that sodium polyethlenesulfonate fitted a
permanent dipole moment saturation in spite of its strong
polyelectrolyte properties.

The above saturation theory was extended by Shah [19], for
the case $\alpha_1 - \alpha_2 < 0$. He investigated bentonite suspensions
[20], and found that the negative birefringence in weak fields
was brought about by permanent dipole along the axis of symmetry
of the bentonite disk, and the positive birefringence in strong
fields was due to ion atmosphere polarization. Holcomb and

Tinoco [21] calculated the saturation effect for the more general model in which the electrical and optical polarizabilities need have no symmetry and the permanent dipole may have any orientation.

D. Effect of Ionic Strength

O'Konski and Zimm [3] found that the electric birefringence of TMV solutions was decreased by increasing the concentration of buffer. O'Konski and Haltner [4] investigated the effect of added simple electrolyte more quantitatively. As the concentration increased, δ/E^2 steadily decreased, and the data could be correlated by the relation

$$\frac{\delta}{E^2} = 27 \times 10^{-9} K^{-0.28}$$

K is the specific conductivity of the solution and is ohm^{-1} cm^{-1}, E is volts cm^{-1}, δ is radians cm^{-1}. The conductivity of the solvent reduces the effect of counterion atmosphere conductance, and this will be discussed in the next section. Besides, the ionic strength affects the thickness of ion atmosphere, and alters the ion atmosphere polarization.

E. Effect of pH

Most polyelectrolytes contain a large number of ionizable acidic and basic groups, so that the pH of the solution would have an important effect on the ionization, and accordingly, on the electric birefringence. The transient behavior of the birefringence of TMV was unchanged by altering pH, that is, the relaxation time remained constant at various pHs; this indicated no change of structure over the range studied. But the birefringence at pH 8.4 was higher than that at pH 7.0, and the latter was higher than that at pH 5.6 under similar conditions. Since increasing the pH increases the TMV charge, which may be computed from titration data, the surface conductivity is higher at higher pH. Accordingly, the variation of Kerr constant was reasonable in terms of a counterion polarization orientation mechanism.

Kirkwood and Shumaker [14], proposed that the dielectric increments of protein solutions may be explained by the migration of protons among the acidic and basic sites on protein. At lower pH values proton transfer would become important and at higher pH values the OH⁻ would contribute. This proton polarization mechanism may be regarded as a special cause of the more general ion transport phenomenon [8].

When orientation arises from a permanent dipole moment, the change of dipole moment due to the change of pH is also important. For collagen solution it was observed that permanent dipole moment decreases and induced dipole moment increases as pH is raised [22].

F.   Effect of Concentration

Generally, in concentrated solutions strong dependences of the birefringence and its relaxation behavior upon concentration are observed. In most cases these are due to macromolecular interaction or association. An analysis of the results on TMV led to the conclusion that end-to-end dimers are present [23]. The result on deoxyribonucleic acid solutions shows that macromolecular interactions exist even in relatively dilute solutions [24, 25]. The results suggest that electric birefringence is useful in studies of molecular interactions. The macromolecular interaction affects the permanent dipole and the ion atmosphere. In collagen solution a decrease of solute concentration reverses the sign of the birefringence [22].

G.   Hydrodynamic Orientation

The production of an orienting torque by the hydrodynamic forces accompanying electrophoresis was considered [3,26,27]. Anisometric bodies might hydrodynamically orient with long axes across the direction of motion [3], but direct evidence for this type of orientation mechanism has not been reported.

III.   IONIC POLARIZATION IN POLYELECTROLYTE

Ion atmosphere polarization is the basic problem in the treatment of electro-optic properties of polyelectrolytes. The concept of surface conductivity was introduced in treating the ion atmosphere polarization by O'Konski and this model was treated mathematically [28,8] using the Maxwell-Wagner polarization theory. The idea of surface conductivity had been introduced by Smoluchowski to discuss the electrophoresis of colloidal particles. A Maxwell-Wagner polarization model had been applied in different ways to discuss the conductivity or dielectric properties of colloidal systems [29-31].

The surface conductivity is the rate of transfer of charge along a surface through the unit length in an electric field with a unit component along the surface perpendicular to a unit length line. If u_i is the mobility of a charge carrier i, the surface conductivity λ is

$$\lambda = \sum_i n_i u_i z_i \tag{1}$$

where n_i is the number of charge carriers i per unit surface area, and z_i is the absolute value of the charge [8]. Calculations of surface conductivity were made on various double-layer models.

For spherical polyelectrolyte of radius a, Laplace's equation was solved under appropriate boundary conditions [8]. The result shows that the effect of the surface conductivity λ is equivalent to an increase of the volume conductivity of spherical polyelectrolyte of radius a, κ_2, by $2\lambda/a$, namely,

$$\kappa_s = \kappa_2 + 2\lambda/a \tag{2}$$

where κ_s is the effective volume conductivity.

The surface contributions κ_a', κ_b', and κ_c to the volume conductivity for a ellipsoid along its semiaxes a, b, and c were approximated by the same considerations similar to those for a sphere [8]. The surface contributions, κ_a', κ_b', κ_c' along the respective axes were found to be

$$\kappa_a' = \frac{C(b,c)\lambda}{\pi bc}$$

$$\kappa_b' = \frac{C(a,c)\lambda}{\pi ac} \tag{3}$$

$$\kappa_c' = \frac{C(a,b)\lambda}{\pi ab}$$

where $C(j,k)$ is circumference of the ellipse with semiaxes j and k, and is approximately equal to $2\pi[(j^2 + k^2)/2]^{1/2}$ for the nearly circular cross sections and approaches $4(j + k)$ for highly eccentric cases. Furthermore, the κ's depend upon position along the axis, but they are nearly constant for interesting cases. Thus, for a slightly oblate or prolate spheroid, the surface contribution can be approximated by that for a sphere, $2\lambda/a$. For oblate or prolate ellipsoid of revolution (b = c) the following equation is adequate in most of the interesting polyelectrolytes.

$$\kappa_a' = 2\lambda/b$$

For prolate ellipsoids of revolution of a >> b = c

$$\kappa_b' = \kappa_c' = \frac{4\lambda}{\pi b} \tag{4}$$

For this oblate ellipsoids (a << b = c)

$$\kappa_b' = \kappa_c' = \frac{4\lambda}{\pi a} \tag{5}$$

For a cylinder of length   2a   and diameter   2b   the axial and transverse surface contribution, κ_a' and κ_b' are

$$\kappa_a' = \frac{2\lambda}{b} \tag{6}$$

$$\kappa_b' = \frac{(a + b)\lambda}{ab} \tag{7}$$

So that, for rods   (a >> b)

$$\kappa_b' = \frac{\lambda}{b}$$

for disks   (a << b)

$$\kappa_b' = \frac{\lambda}{a}$$

In this way the surface conductivity was converted into the equivalent volume conductivity, with the rigid polyelectrolytes regarded as rigid conducting particles. Here only the ellipsoidal polyelectrolytes will be considered, because most of the interesting polyelectrolytes may be approximated by an ellipsoid. The theory of the orientation effect of an electric field on a rigid conducting ellipsoid carrying a permanent dipole moment in a conducting medium was developed recently by O'Konski and Krause [11].

The result of optical birefringence Δn obtained for weak electric fields are as follows,

$$
\begin{aligned}
\Delta n &= \frac{2\pi}{15n} C_v E^2 [(g_a - g_b)(P_a - P_b + Q_{ab}) \\
&\quad + (g_b - g_c)(P_b - P_c + Q_{bc}) \\
&\quad + (g_c - g_a)(P_c - P_a + Q_{ca})] \\
&= K_{sp} C_v n E^2
\end{aligned} \tag{8}
$$

Definitions of the symbols are given in Chap. 3.

For rodlike macromolecules of very high axial ratio, provided that $\kappa_a/\kappa_b \gg 1/A_a$ which corresponds to very high ionic conductivities, it was shown that

$$K_{sp} = \frac{\varepsilon v(g_n - g_b)}{30n^2 kTA_a} \qquad (9)$$

For ellipsoids of revolution of arbitrary axial ratio but without a dipole moment

$$K_{sp} = \frac{\varepsilon v(g_a - g_b)}{30n^2 kT} \left[\left(\frac{\kappa_a}{\kappa} - \frac{\varepsilon_a}{\varepsilon}\right) B_a^2 + \left(\frac{\kappa_a}{\kappa} - 1\right) B_a \right.$$
$$\left. - \left(\frac{\kappa_b}{\kappa} - \frac{\varepsilon_b}{\varepsilon}\right) B_b^2 + \left(\frac{\kappa_b}{\kappa} - 1\right) B_b \right] \qquad (10)$$

Other special cases also were treated [11].

The above theory applies only when the ionic atmosphere relaxation time is short compared to the orientation relaxation time. With TMV dispersion occurs above 10 kHz, and the ion atmosphere polarization cannot achieve steady state conduction. Hence the above theory is expected to apply below 10 kHz. This fact should be kept in mind in birefringence dynamics studies of relatively small molecules.

The experimental value of K_{sp} for TMV in 1.5×10^{-4} M phosphate buffer was 8.5×10^{-4} cgs unit [4]. If TMV is treated as insulating ellipsoid of revolution Peterlin-Stuart theory gave a K_{sp} of 1.6×10^{-5} cgs unit. This difference should be attributed to counterion polarization. If TMV is regarded as a very good conductor, the above theory gives 34×10^{-4} cgs unit. Calculations for finite conductivity were expected to give better agreement. From those data O'Konski and Krause made a quantitative test of the more recent theory [11], and the prediction of the Kerr constant from estimated counterion mobilities was found to be reasonable.

IV. STUDIES ON POLYELECTROLYTES

The electro-optic properties of many polyelectrolytes, such as biopolymers, synthetic polymers, and colloid systems have been investigated. In this section we will review the main results.

A. Tobacco Mosaic Virus

Since TMV is an excellent material for model studies of electric birefringence, many measurements have been carried out on it. Lauffer [32], measured the birefringence of TMV in an ac field of 60 Hz, and concluded that TMV rodlike molecules are oriented with long axis parallel to the electric field as the birefringence was positive. He inferred a permanent moment was responsible for the orientation. O'Konski and Zimm [3] applied the pulse technique, and suggested that the orienting mechanism is an induced polarization due to the counterion atmosphere. The conclusion against a dipole mechanism was obtained from the frequency dependence of the birefringence [33].

O'Konski and Haltner [23,4], carried out a systematic study on TMV which was described in Sec. II of this chapter. O'Konski and Pytkowicz [34], found by using the reversing pulse technique that the Holmes Rib Grass strain of TMV showed a large permanent dipole moment of about 10^4 to 10^5 D. This was in striking contrast to the previous results, which were obtained for common strain. The difference is not explained.

Allen and Van Holde [35] studied the dichroism of TMV in pulsed electric fields. Saturation of the orientation was easily achieved, and the saturation field was higher for concentrated solutions. This is probably due to the fact that polarization of the medium reduced the effective field, and this polarization can become large at high concentrations. On the other hand, the small decrease in the dichroism with concentration is hard to explain. They obtained an electric polarizability, which was of the

same order of magnitude as that obtained from the electric
birefringence measurement of O'Konski et al. [5].

B. Polynucleotides

Some electro-optic studies on the aqueous solutions of
deoxyribonucleic acid (DNA) and synthetic polynucleotides were
reported. DNA is not a rigid polyelectrolyte, and the experimental results are more complicated than TMV. Benoit [15], and
Norman and Field [36], reported the experimental results on DNA
and obtained the length of molecules, but little attention was
paid to polyelectrolyte properties.

Stellwagen and others [24,25] made investigations on calf
thymus DNA in dulute aqueous solution, variables studied being DNA
concentration, buffer concentration, sonication, and heat treatment. The sign of the birefringence was negative. Since the flow
birefringence of DNA is also negative, the DNA molecules must
orient with their long axes parallel to the electric field. The
negative sign of the birefringence is due to the large
polarizability of bases which is perpendicular to long axis as expected from Watson and Crick model. They studied also the behavior of the birefringence in the reversing field. The transient
effects of opposite sign were observed upon field reversal. This
reversal may be due to the transverse electric polarizability.
The field dependence of the birefringence did not obey the Kerr
law even in weak fields, and in some cases the birefringence increased approximately linearly with field strength. This interesting behavior also is observed for other polyelectrolytes [24,
37]. There are many complicated and interesting observations regarding the DNA solutions (see Chap. 18) and further studies are
necessary.

It was found that the birefringence of aqueous solutions of
sodium DNA is anomalous when electric fields of high intensity
are applied [38]. It was proposed that the electric field

probably causes aggregation of the macromolecules and then
produces a structural transition concomitant with the electric
field. Field-induced transitions are discussed in Chap. 3.

    Golub [39], considered the method for estimation of chain
macromolecule rigidity from the electric birefringence, and
applied it to the solution of DNA. According to Beck and Hermans
[40], and Bueche [41], there is in high polymer solutions a region of dielectric dispersion which is due to the movements of
segments of the macromolecular chain backbone, and the dispersion
in this region can be described by Debye formula with a single
relaxation time, which is determined by the chain segment and is
independent of the molecular weight. The relaxation time found
experimentally would then characterize the rigidity of the
macromolecules. With some nonionic polymers they found that the
relaxation times were the same even for samples of various molecular weight. When DNA is dissolved in water, its rigidity is in
part a result of repulsions between like charges of the polyanion.
The relaxation time of water solutions of DNA were found to be
much higher than those containing salt.

    Jakabhazy and Fleming [18], studied the electric birefrin-
gence of synthetic polynucleotides, poly A and poly U, and their
interaction. They used the electric pulses consisting of 25 kHz
sine wave fields to avoid electrophoresis and decomposition at the
electrodes. This results support the conclusion that ion atmosphere polarization is responsible for electro-optic effects in
solutions of polynucleotides. Mg^{2+} binds to the phosphate part of
the backbone, and in the presence of Mg^{2+} the kinetics of the
double-stranded helix formation of poly(A + U) is second order and
the Kerr effect of poly A and poly U is sharply reduced by increasing the ratio of magnesium to phosphate. Mg^{2+} also strongly
decreases the relaxation times. Houssier and Fredericq [42]
measured the electric birefringence and dichroism of DNA and
nucleoproteins. A very sharp influence of the ionic strength was
observed, the dichroism being decreased with increasing ionic

strength. They also reported the electro-optic properties of
gel-forming deoxyribonucleohistone [43]. It was observed that
the relaxation times increased sharply with decreasing field
strength, and both the birefringence and the dichroism were
sharply diminished by increase in the ionic strength and by decrease in the pH from 7 to 5.5. The relaxation times were only
slightly lowered by enzymic degradation and were much less influenced by the field strength and by the ionic strength. These
results are consistent with the presence of aggregation in the
gel-like solutions which could be disordered by increasing the
ionic strength and decreasing the pH. Shirai [44] measured the
ultraviolet electric dichroism of DNA and poly A in the strong
electric fields, and obtained values concerning electric properties.

C. Collagen

The electric birefringence of collagen from rat tail tendon
was measured over a wide range of field strength by Yoshioka and
O'Konski [45]. The solvent was dilute acetic acid. Very pronounced saturation of the electric birefringence was observed,
and from the field dependence of the birefringence the permanent
dipole moment and the anisotropy of electric polarizability were
determined. The contribution of the former to the Kerr constant
was found to be twice as large as that of the latter. The same
conclusion was obtained from the initial slope of the rise curves
of the birefringence at low fields. It is interesting that both
contributions are the same order of magnitude. The permanent
dipole moment was 1.5×10^4 D, and the anisotropy of electric
polarizability was about 3×10^{-15} cm^3. The ion atmosphere
polarizability decreased with the addition of salt, qualitatively
in agreement with the predictions of O'Konski and Krause's
theory [8,11]. Kahn and Witnauer [46], measured the electric

birefringence of calf skin corium collagen dissolved in citrate
buffer in the acid pH range. At high concentrations of collagen,
they observed an anomalous electric birefringence patterns which
they considered to be indicative to time-dependent variations in
permanent and induced moments.

D. Other Proteins

Proteins have acidic and basic groups, and are
polyelectrolytes. The electro-optic behavior of some proteins
other than collagen were studied. However, their orientation
mechanism was permanent dipole polarization. For example, the
studies of bovine serum albumin by Moser et al. [47] showed that
the protons of isoionic protein were not sufficiently mobile to
contribute a large ionic polarization or a fluctuation polarization. Thus their electro-optic properties as polyelectrolytes
have not been studied.

Tinoco [48], studied the bovine fibrinogen in the pH range
from 6 to 10 in a urea-water-glycerol solvent. Haschemeyer and
Tinoco [49], investigated bovine fibrinogen in the pH ranges
4.0-5.0 and 7.0-9.0. A permanent dipole moment along the transverse axis in the low pH region was found, and one along the
symmetry axis was found in the high pH region. These results
were explained in terms of a distribution of charged groups in
the molecule.

Krause and O'Konski [50,51], measured the electric birefringence of small globular proteins by using fast intense pulses
from a delay-line pulse generator. They studied bovine serum
albumen, γ-globuline, ovalbumin and others. The effects of pH
and ionic strength were studied for bovine serum albumin. The
orientation was mainly due to a permanent dipole. Riddiford and
Jennings [52] studied aqueous solutions of bovine serum albumin
at various pH. At the pH lower than the isoelectric value 5.1
the relaxation time fell and this was accompanied by a large

increase in the electric anisotropy. They also studied solutions
of ovalbumin, bovine γ-globulin, and β-lactoglobulin [53].   The
decrease of relaxation time of γ-globulin at lower pH was observed. Orttung [54], measured the Kerr constant of met- and
oxyhemoglobin, and concluded that it showed a permanent dipole
moment.

E.   Synthetic Polyelectrolytes

Stellwagen [24], made electro-optic investigation on sodium
polyphosphate (NaPP). The sign of the birefringence was positive.
Since the flow birefringence of NaPP is positive, the NaPP molecules must orient with their long axes parallel to the electric
field. The field dependence of the birefringence of NaPP did not
obey Kerr law even in low field strengths, the limiting slopes of
$\log \delta$ vs. $\log E$ curves fell between 1.0 and 2.2. The magnitude
of the deviations of NaPP from the Kerr law corresponded roughly
to the lengths of the molecule. She obtained the optical anisotropy factor of about 10. This value is in agreement with that
calculated from the birefringence of oriented fibers. Specific
Kerr constants of various samples were $2\text{-}5 \times 10^{-5}$ esu. Since
these samples did not obey the Kerr law, the values must be
considered lower limit of the true values. Since the birefringence of NaPP did not obey the Kerr law, the mechanism of polarization could not be determined from the saturation curve of the
birefringence. However, if the electrical factor is attributed
to the polarizability, the electrical anisotropy factors are
$0.4\text{-}1 \times 10^{-14}$ cm^3 for various samples. If the electrical factor
is attributed to dipole moment, the dipole moments obtained for
various samples are $4\text{-}6 \times 10^3$ D.

Yoshioka and O'Konski [55], studied the electric birefringence of sodium polyethylenesulfonate in water. The field dependence of a low molecular weight fraction followed the Kerr law
quite closely. A high molecular weight fraction had a much

larger Kerr constant and showed saturation in a field of 15 kV/cm.
The specific Kerr constant increased with decreasing polyelectrolyte concentration in pure water like other properties of dilute
solutions of linear flexible polyelectrolytes such as the reduced
viscosity and the specific dielectric increment. On the other
hand, the specific Kerr constant decreased on isoionic dilution.
The saturation effect was the more marked, the lower the polyelectrolyte concentration. The saturation behavior resembled that of
permanent dipole orientation, but this mechanism was not supported
by the buildup of the birefringence. The behavior of the electro-
optic relaxation times with added salt and with concentration is a
sensitive indication of the flexibility of a charged macromolecule
or a polyelectrolyte.

The electric birefringence of potassium polystyrenesulfonate
(KPSS) in water and dioxane-water mixtures was studied by
H. Nakayama and K. Yoshioka [56]. It obeyed the Kerr law in low
fields. The saturation of the birefringence was observed at
higher field strength, and it is the more pronounced, the lower
the polyelectrolyte concentration. The specific Kerr constant decreased with increasing polyelectrolyte concentration and with
increasing dioxane content. The electric birefringence Δn of
KPSS in glycerol-water and ethylene glycol-water mixtures is related to the field strength E by the equation, $\Delta n/c = KE^n$,
where c is the polyelectrolyte concentration. The exponent n
changes from 1.05 to 1.70. The deviation from the Kerr law is the
more marked, the larger the fraction of glycerol. This relation
is similar to that found by Stellwagen [24]. The similar birefringence behavior was also observed in KPSS in water-dioxane mixtures [57]. The deviation from Kerr law observed in solutions of
DNA and KPSS is an interesting problem.

F. Colloidal Electrolytes

The electric birefringence of V_2O_5 sols was observed by many authors [1,2], and positive birefringences was found. The electric birefringence of various metallic and nonmetallic particles in suspension was also studied [58]. Metallic sols generally gave negative birefringence, whereas in nonmetallic sol, both signs occurred. For all of these systems the counterion atmospheres probably exert large effects.

Shah et al. [20] investigated the various monodisperse bentonite suspensions. Suspensions of various concentrations and particle sizes were employed. They concluded that the negative birefringence in weak fields is brought about by a permanent dipole along the axis of symmetry of the bentonite disks. The permanent dipole orientation is perpendicular to the strong field orientation, which is predominantly governed by the ion atmosphere polarization. Experimental results with square wave and ac fields supported this model.

Stoylov and Petkanchin [59] studied the influence of electrolytes on the electric polarizability for TMV, palygorskite (a type of clay with strongly elongated rodlike particles) and bentonite by the electric light scattering. KCL was used as electrolyte. The electric polarizability of bentonite was not influenced by the electrolyte, but the other two substances were strongly decreased. The effect of ionic strength for TMV and polygorskite is lower for frequencies of the applied electric field where relaxation of the electric polarizability was observed, and for the highest frequencies palygorskite is not influenced by the ionic strength at all.

The permanent dipole moment of bentonite is interesting. In its contribution to the Kerr effect it compares with the induced dipole moment. The values of the permanent dipole moment obtained by different authors [59-66] differ considerably, $0.6-9.0 \times 10^5$ D, and even the direction is different. The form of bentonite particle is disk-like, and some authors reported the permanent dipole moments directed along the short axis [59-65], but others report those directed along the long axis [66].

The scattering of the values of the dipole moment obtained by different authors are probably connected with the different types of bentonite used or with different electrolyte composition. The difference of direction of the dipole moment was interpreted by some authors as an indication of the presence of two types of particle--one with high electric polarizability along the long axis and the other with permanent dipole moment directed along the short axis [60,62]. But it seems more reasonable to explain the result in terms of only one type of particle possessing electric polarizability along the long axis and permanent dipole moment along the short axis [20,61].

V.   POLYELECTROLYTE-DYE COMPLEXES

A very interesting problem regarding the polyelectrolyte-dye complexes is the structural consideration in the interaction of DNA and dyes. The combination in solution of DNA with small amounts of dyes such as acridine, proflavin, or acridine orange was found to enhance the viscosity and decrease the sedimentation coefficient of DNA [67]. Characteristic changes, which suggest considerable modification of the usual helical structure of DNA, are found in the x-ray diffraction patterns of fibers of the complex with proflavin. It was inferred that dye molecules are intercalated between adjacent nucleotide-pair layers by extension and unwinding of the deoxyribose-phosphate backbone [67]. How-

ever, this model is open to question, and has not adequately
explained certain experimental results [68,69].

Stellwagen [24], studied the birefringence of DNA-dye,
polynucleotides-dye, and polyphosphate-dye complexes. Bradley et
al. [68], studied the birefringence and dichroism of DNA-dye and
polynucleotide-dye complexes. Generally, polyelectrolytes orient
with their long axes parallel to the electric field, so that if
complexes have only randomly oriented chromphores, they show positive birefringence. The dyes, acridine orange (AO) and methylene
blue (MB), were used. The DNA-AO, poly U-AO, and poly G-AO as
well as DNA-MB complexes exhibited negative birefringence and
perpendicular dichroism. Since the polymer molecules are aligned
parallel to the electric field, the AO molecules must lie with
their planes more or less-perpendicular to the field. The magnitudes of the birefringence of AO complexes are generally greater
than those of the polymers themselves. Poly A-AO complex showed
positive birefringence and parallel dichroism, indicating that the
dye molecules are oriented with their long axes parallel to the
field. The sign of the birefringence of MB-DNA complex, was negative. The magnitude of the birefringence was considerably less
than that of DNA alone. A cutoff filter was used to minimize the
complication of dichroism signals due to the absorption band of
the complexes in these measurements.

Houssier et al. [42], measured the electric dichroism of the
complex DNA-proflavin. It showed negative dichroism and the
dichroic ratio D was about 1.6 at 6 kV/cm. The D of
deoxyribonucleoprotein (DNPr) was much lower than that of DNA.
The ionic strength had a strong influence. D increased with DNPr
concentration and reached saturation for a molar ratio
M(DNPr)/M(proflavin) of about 12, correlated with the maximum displacement of the maximum of the absorption band (λ_{max} = 441.5 nm
for proflavin; λ_{max} = 458 nm for the complex). They observed an
increase of the electric birefringence for complex compared with
DNPr alone. The above data are consistent with a mainly perpendicular orientation of the planes of the base rings with respect to

the long axis of both DNA and DNPr particles. This result may be
in agreement with the insertion or intercalation scheme [67].

Houssier et al. [13] studied the interaction between the
gel-forming deoxyribonucleohistone from calf thymus and proflavin
by means of the electro-optic method in media of very low ionic
strength. The complexes displayed a negative dichroism in the
visible range, due entirely to the dye. The dichroic ratio D
increased with decreasing numbers of dye molecules per atom of
phosphorus, r, and reached a maximum value of about 1.85 for
$r \leq 0.10$, approximately constant in the visible region at the
field strength of 13 kV/cm. The relaxation times were not noticeably affected by the interaction except for $r > 0.10$ where a decrease of relaxation time was observed. The results appear to be
consistent with an intercalation of the proflavin cation between
adjacent nucleotide pairs, for at most one proflavin molecule per
five nucleotide pairs in the gel-forming deoxyribonucleohistone.
For higher proflavin contents, the dye molecules would be exter-
nally attached.

In the vicinity of an electronic absorption, the Kerr constant exhibits anomalous dispersion, and the form of this dispersion curve was related to the direction of the transition moment
for the particular electronic transition. Thus the measurement of
dispersion of the Kerr constant is useful for the study on
polyelectrolyte-dye complexes. Powers [12] measured the Kerr
effect in the vicinity of an electronic absorption for the
acridine orange-polyglutamic acid complex in the solvent
dimethylformamide (DMF). The much higher conductivity of water
solutions made it impossible to measure the birefringence with
sufficient accuracy in aqueous solutions. The dye molecule must
be bound with its long axis approximately parallel to the polymer
axis, because the complex shows positive dispersion and the free
dye has the long-axis polarization at the long-wavelength absorption (5100 Å). Powers [70] also investigated the structure of
the proflavine-polyglutamic acid complex by measuring the

dispersion of the Kerr constant. A positive dispersion curve was found, indicating that the dye is bound with its long axis parallel to the long axis of the polymer. The absence of optical activity is ascribed to the lack of strong dye-dye interactions. These interactions are present in other acridine dyes.

Soda and Yoshioka [71] studied the electric dichroism of crystal violet-DNA and crystal violet-potassium polystyrenesulfonate systems in aqueous solution. The results indicated that the molecular plane of crystal violet is parallel to that of polystyrenesulfonate ion in the large excess of polyelectrolytes.

VI.   CONCLUDING REMARKS

The preceding sections show that the electric birefringence and dichroism are useful methods for studies of polyelectrolyte solutions. These methods supply information about the electrical properties, as well as the optical and hydrodynamic properties. The electrical properties of macromolecules also are studied by the dielectric dispersion, but for measurement of dielectric dispersion relatively concentrated and low conductivity solutions are needed. Concentrated polyelectrolyte solutions are conductive, and it is very difficult to measure their dielectric dispersion. The electro-optic method using the pulse technique has largely overcome this difficulty. Furthermore, this method extends our knowledge about geometrical and optical properties. Since the electro-optic method is applicable to the more conductive solutions, the dependence of various electro-optic properties of solutions on concentration, pH, and ionic strength can be measured. It is also the merit of the electro-optic method to know the polarization mechanism by the saturation curve or by the reversing pulse technique.

There are many interesting and unsolved problems concerning the electro-optic or electric properties of polyelectrolyte solutions. Some polyelectrolyte solutions do not obey the Kerr law even at low fields. DNA solutions shows unusual transients on reversing the pulse. Polyelectrolyte solutions generally have large dielectric constants. The explanation of these experimental results may be important for a general understanding of polyelectrolyte solutions.

Complexes of polyelectrolytes with dyes are also interesting substances for further electro-optic studies. Since orientation of chromophores in the complexes can be determined by the electric birefringence or dichroism measurements, these methods are powerful for investigating the complexes and for elucidating their structures.

REFERENCES

1. C. Bergholm and Y. Bjornstahl, Phys. Z., 21, 137 (1920).

2. J. Errera, J. Th. G. Overbeek, and H. Sack, L. Chim. Phys., 32, 681 (1935).

3. C. T. O'Konski and B. H. Zimm, Science, 111, 113 (1950).

4. C. T. O'Konski and A. J. Haltner, J. Am. Chem. Soc., 79, 5634 (1957).

5. C. T. O'Konski, K. Yoshioka, and W. H. Orttung, J. Phys. Chem., 63, 1558 (1959).

6. M. A. Lauffer, J. Phys. Chem., 42, 935 (1938).

7. K. Yoshioka and C. T. O'Konski, J. Polymer Sci., 6, 421 (1968).

8. C. T. O'Konski, J. Phys. Chem., 64, 605 (1960).

9. G. Schwarz, J. Phys. Chem., 66, 2636 (1962).

10. M. Schurr, J. Phys. Chem., 68, 2407 (1964).

11. C. T. O'Konski and S. Krause, J. Phys. Chem., 74, 3243 (1970).

12. J. C. Powers, Jr., J. Am. Chem. Soc., 89, 1780 (1967).

13. C. Houssier and E. Fredericq, Biochim. Biophys. Acta, 120, 434 (1966).

14. J. G. Kirkwood and J. B. Shumaker, Proc. Nat. Acad. Sci., 38, 855 (1952).

15. H. Benoit, Ann. Phys., 6, 561 (1951).

16. I. Tinoco, Jr., J. Am. Chem. Soc., 77, 4486 (1955).

17. I. Tinoco, Jr., J. K. Yamaoka, J. Phys. Chem., 63, 423 (1959).

18. S. Z. Jakabhazy and S. W. Fleming, Biopolymers, 3, 793 (1966).

19. M. J. Shah, J. Phys. Chem., 67, 2215 (1963).

20. M. J. Shah, D. C. Thompson and C. M. Hart, J. Phys. Chem., 67, 1170 (1963).

21. D. N. Holcomb and I. Tinoco, Jr., J. Phys. Chem., 67, 2691 (1963).

22. L. D. Kahn and L. P. Witnauer, Biochim. Biophys. Acta, 243, 398 (1971).

23. C. T. O'Konski and A. J. Haltner, J. Am. Chem. Soc., 78, 3604 (1956).

24. N. C. Stellwagen, Ph.D. Thesis, University of California, Berkeley, Calif., 1967.

25. C. T. O'Konski, J. C. Stellwagen, and M. Shirai, Biophys. J. (in press).

26. W. Heller, Rev. Mod. Phys., 14, 890 (1942).

27. K. J. Mysels, J. Chem. Phys., 21, 201 (1953).

28. C. T. O'Konski, J. Chem. Phys., 23, 1959 (1955).

29. J. B. Miles and H. P. Robertson, Phys. Rev., 40, 583 (1932).

30. J. J. Bikerman, J. Chim. Phys., 32, 285 (1935).

31. H. Fricke, J. Phys. Chem., 57, 934 (1953).

32. M. A. Lauffer, J. Am. Chem. Soc., 61, 2412 (1939).

33. H. Benoit, J. Chim. Phys., 49, 517 (1952).

34. C. T. O'Konski and R. M. Pytkowicz, J. Chim. Phys., 79, 4815 (1957).

35. F. S. Allen and K. E. Van Holde, Biopolymers, 10, 865 (1971).

36. A. Norman and J. A. Field, Jr., Arch. Biochem. Biophys., 71, 170 (1957).

37. H. Nakayama and K. Yoshioka, J. Polymer Sci., 3, 813 (1965).

38. C. T. O'Konski and C. Stellwagen, Biophys. J., 5, 607 (1965).

39. E. I. Golub, Biopolymers, 2, 113 (1964).

40. L. K. H. Van Beck and J. J. Hermans, J. Polymer Sic., 23, 211 (1957).

41. F. Bueche, J. Polymer Sci., 54, 597 (1961).

42. C. Houssier and E. Fredericq, Biochim. Biophys. Acta, 88, 450 (1964).

43. C. Houssier and E. Fredericq, Biochim. Biophys. Acta, 120, 113 (1966).

44. M. Shirai, J. Chem. Soc. Japan, 86, 1115 (1965).

45. K. Yoshioka and C. T. O'Konski, Biopolymers, 4, 499 (1966).

46. L. D. Kahn and L. P. Witnauer, Biochim. Biophys. Acta, 243, 388 (1971).

47. P. Moser, P. G. Squire, and C. T. O'Konski, J. Phys. Chem., 70, 744 (1966).

48. I. Tinoco, Jr., J. Am. Chem. Soc., 77, 3476 (1955).

49. A. E. V. Haschemeyer and I. Tinoco, Jr., Biochemistry, 1, 996 (1962).

50. S. Krause and C. T. O'Konski, J. Am. Chem. Soc., 81, 5082 (1959).

51. S. Krause and C. T. O'Konski, Biopolymers, 1, 503 (1963).

52. C. L. Riddiford and B. R. Jennings, J. Am. Chem. Soc., 88, 4359 (1966).

53. C. L. Riddiford and B. R. Jennings, Biopolymers, 5, 757 (1967).

54. W. H. Orttung, J. Am. Chem. Soc., 87, 924 (1965).

55. K. Yoshioka and C. T. O'Konski, J. Polymer Sci., 6, 421 (1968).

56. H. Nakayama and K. Yoshioka, J. Polymer Sci., 3, 813 (1965).

57. H. Nakayama and K. Yoshioka, J. Chem. Soc. Japan, 85, 177 (1964).

58. Y. Bjornstahl, Phil. Mag., 2, 701 (1926).

59. S. Stoylov and I. Petkanchin, Izv. Fizikokhim., Bulgar. Akad. Nauk., 5, 73 (1965).

60. C. Wippler, J. Chim. Phys., 53, 328 (1956).

61. M. J. Shah and C. M. Hart, IBMJ Res. Dev., 7, 44 (1963).

62. B. R. Jennings and H. G. Jerrard, J. Chem. Phys., 42, 511 (1965).

63. S. Stoylov, Izv. Inst. Fizikokhim., Bulgar, Adak. Nauk, 6, 79 (1967).

64. S. Stoylov, S. Sokerov, I. Petkanchin, and N. Ibroshev, Dokl. Akad. Nauk SSSR, 180, 1165 (1968).

65. G. B. Thurston and D. I. Bowling, J. Colloid Interface Sci., 30, 34 (1969).

66. N. A. Tolstoi, A. A. Spartakov, and A. A. Trusov, Issled. Obl. Poverkh Sil, Sb. Dokl. Konf., 3rd, Moscow, 1967, p. 56.

67. L. S. Lerman, J. Mol. Biol., 3, 18 (1961).

68. D. F. Bradley, M. C. Stellwagen, C. T. O'Konski, and C. M. Paulson, Biopolymers, 11, 645 (1972).

69. J. Paolotti and J. Le Pecq, J. Mol. Biol., 59, 43 (1971).

70. J. C. Powers, Jr., IUPAC meeting Boston, 1971, p. 170.

71. T. Soda and K. Yoshioka, J. Chem. Soc. Japan, 86, 1019 (1965); 87, 1326 (1966).

Chapter 20

ELECTRO-OPTICS OF NUCLEOPROTEINS AND VIRUSES

Marcos F. Maestre

Space Science Laboratory
University of California
Berkeley, California

I.   INTRODUCTION

The study of viral particles and nucleoproteins by electro-
optic methods has provided useful information with respect to the
internal organization of the nucleic acids of these biological
structures. In particular the viruses, which are fully biologically competent, are extremely homogeneous in size shape and chemical properties allowing the measured electro-optical properties of

the bulk of a total population to be assigned to each member of
the population.

II.   VIRAL STRUCTURES

Viral structures which have been studied by electro-optic
techniques include several strains of tobacco mosaic virus (TMV),
the bacteriophages T2, T4, and fl [1-13].

Information about the arrangement of the in situ structures
of DNA or RNA can be obtained in terms of the anisotropy of the
optical polarizability tensor, as a function of wavelength and
with respect to a fixed frame of reference.

Both electric dichroism and birefringence techniques produce
orientation of these highly anisotropic viral structures and thus
provide a direction with respect to which anisotropies in the
electric and optical tensors of the nucleic acid chromophores can
be measured. This information can then be used in combination
with other physical and chemical studies to construct a model of
the internal organizations of the nucleic acids in viruses,
nucleohistones, and other biological structures.

A. Morphology

The size and shape of nucleoproteins and viruses can be
determined only to the degree in which good preparations for
electron microscope, x-ray crystallography, and hydrodynamic
measurement studies are obtained. Some of the viral structures
have been studied to an extraordinary degree of detail. Probably
the most detailed analysis of structure is the case of TMV which
has been intensively studied by a variety of physical and chemical
techniques since the 1930s. (See reviews by Stoylov [13],
Fraenkel-Conrat [14], and Caspar [15].) This plant virus is the
ideal biological subject for study by electro-optic methods. It

has a large axial ratio (a/b = 20), good chemical stability and
homogeneity and it is easy to obtain in reasonably large quantities. The disassembly of TMV by cold glacial acetic acid into its
component protein subunits by Fraenkel-Conrat and Williams [16],
and its subsequent reassembly into a viable virus has been one of
the key experiments toward the understanding of the organization
of this virus.

The comparison of the radial electron distribution of intact
TMV with the reaggregated RNA-free rod-shaped protein copied from
the virus has given evidence for the location of the nucleic acid
of the virus [15].

TMV has a mean diameter of 150 Å with the maximum protusion
of the subunits giving a maximum diameter of 180 Å. It consists
of 130 turns of a helix with the number of units per turn given as
49 subunits/3 turns of helix. The pitch of the helix is 23 Å [15,
14]. The RNA is coiled in a helix of 80 Å diameter and with the
same pitch. The RNA is bound to the protein in a ratio of three
nucleotides per protein subunit -- probably to two arginine and
one lysine residue [14].

With these facts it is possible to calculate the total length
of the virus from the molecular weight of the RNA. For the RNA
weight of 2.05×10^6 daltons the length computed is 3000 A which
agrees with the values obtained from electron micrographs of the
virus.

Tobacco Mosaic Virus is composed of 2130 protein subunits,
6390 ribonucleotides with a particle weight of 29.4×10^6 daltons.
The protein subunit is composed of a single chain of 158 amino
acids. The kind and distribution of these amino acids among the
various related strains of TMV are reported in Chapter V of
Fraenkel-Conrat's excellent book Chemistry and Biology of Viruses
[14].

A very large amount of experimental data on the chemical and
physical properties of TMV has accumulated over the years. Most
pertinent to this discussion is the flow birefringence and

dichroism data and, of course, the electro-optic measurements on
this virus.

The T2 and T4 bacteriophages and coliphages have been
exhaustively studied by hydrodynamic and electron-microscopic
methods. Their biological properties are well discussed in
Stent's excellent book on the molecular biology of viruses [17].
Cummings and Kozloff [18,19] and Cummings and Wanko [20] have presented detailed morphological dimensions from electron microscope
studies for T2 bacteriophage. These are relatively complex
biological structures. They have hexagonal bipyrmidal heads, tail
sheath structure with a tail core inside, tail plate with six tail
spikes and six tail fibers attached [21,18,19].

T2 has a well-defined sedimentation constant which undergoes
a transition as the solvent conditions are altered [22]. The slow
form of the virus (S = 700) is changed into a fast form
(S = 1000) by a changing of the temperature, or increase of Ca^{2+}
and decrease of pH. Below pH 5.8, only the fast form was observed
at temperatures $T \leq 30^{\circ}C$ in phosphate buffers of ionic strength
0.1 M. As the pH was increased the transition from fast to slow
sedimenting form occurred at lower temperatures e.g. at pH 6.25,
$T_{transition} = 20.1^{\circ}C$; at pH 7.34, $T_{transition} = 16.4^{\circ}C$ and at pH
7.54 only the slow sedimenting form appears. The presence of Ca^{2+}
ion appears to maintain the virus in the fast sedimenting form at
all temperatures [19]. T2 phage can easily be dissociated into
intact DNA and its protein component, the capsid or ghost, by
osmotic shock [23]. It is possible to completely purify the
capsid or coat protein by DNAase digestion differential centrifu-
gation and CsCl banding [24,25,26]. The particle weight for
these viruses has been determined very accurately recently by
Dubin et al. [27]. It is for the whole intact particle
$192.5 \pm 6.6 \times 10^{6}$ daltons. The diffusion coefficient for the
virus is $D_{20,w} = 0.295 \times 10^{-7}$ cm^{2}/sec determined by beat laser
spectroscopy [27]. The best values for the structural dimensions
of these viruses are presented in Table 1. Figure 1 presents a

TABLE 1

Structural Dimensions of T2 Phage and Ghost

	Dimensions ($\overset{\circ}{A}$)
Head height	1190 ± 30[a]
T2 head width	800 ± 20[a]
T2 ghost head height	1230 ± 50[a]
T2 ghost head width	860 ± 50[a]
Tail length	1000-1100[b]
Tail fiber length	1300[b]
Normal virus length	2190-2290, axial ratio a/b = 2.74-2.86
Normal ghost length	2230, a/b = 2.6-2.7
Extended tail fibers virus length	3490-3590, a/b = 4.4-4.5
Extended tail fibers ghost length	3530-3630, a/b = 4.1-4.2

[a]Data of Cummings and Kozloff [18].

[b]Data of Brenner et al. [21].

generalized schematic drawing of the structure of T2 or T4 bacteriophage.

T4 is structurally even more interesting than T2 since its amber mutants (which contain a nonsense triplet) when reproducing on restrictive host produce a variety of morphologically defective progeny which will provide a fertile field for electro-optic studies (see Stent [17] and Fraenkel-Conrat [14]. Of particular interest is the amber mutant of T4 (X4E) which is mutated in genes 34, 35, 37, and 38 which code for a part of the tail fibers. When injected in E. coli B, a restrictive host, particles are formed which are similar to complete phage particles except they lack the tail fibers [28]. These tail fibers are essential to the attachment of the T2, T4, and T6 phage to the cell wall of the host, E. coli. In the hydrodynamic behavior of the virus it may be important whether the tail fibers are extended or not. At low

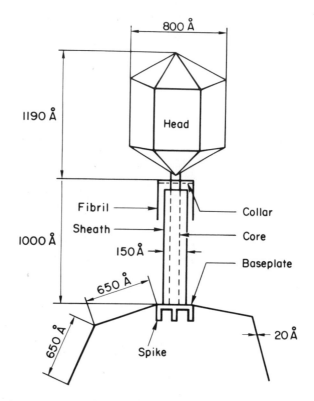

FIG. 1.   A schematic drawing of the main morphological
structures of T2 and T4 bacteriophages.   The DNA is found inside
the head structure.

pH (<5) the fibers of T2, T4, and T6 are folded flat parallel to
the sheath.   At higher pH these fibers extend allowing the absorption of the virus to the host cell wall.   However, there is a
strain of T4, T4B, which needs the presence of tryptophan, tyrosine or phenylalanine as cofactors for the extension of its tail
fibers.   Thus, both T4B and XL4 mutants of T4 phage have been
used for comparative measurements of the hydrodynamic behavior as
a function of tail fiber extension.

The T odd bacteriophages have not been studied so extensively
as the T-even group.   Electron-microscope studies by Brenner and
Horne [29], have given the main shape and size.   The bacteriophage

T5 has an icosahedrally shaped head of 650 Å in diameter with a flexible tail structure of 1700 Å in length by 100 Å diameter [30]. The particle weight of this phage was determined by Dubin et al. [27], to be 109.2×10^6 daltons, approximately half of the T-even phages. The T7 bacteriophage is smaller yet, with an icosahedral head of 470 Å in diameter with a very short tail, 150 Å in length by 100 Å diameter. Its particle weight is 50.4×10^6 daltons [27].

Until very recently, few studies have been done on the biophysical properties of T5, T7 in solution. Camerini-Otero et al. [78], have determined the diffusion coefficient by laser scattering for T7 bacteriophage to be $D_{20,w} = 0.644 \pm 0.007 \times 10^{-7}$ cm^2 sec^{-1}. Using the Einstein-Stokes equations, these authors computed the hydrodynamic radius to be 333 ± 3 Å in reasonable agreement with previously reported values from low-angle x-ray scattering, light scattering, and electron-microscope studies. The hydrodynamic diffusion coefficient for T5 bacterio-phage was determined to be $D_{20,w} = 0.397 \times 10^{-7}$ cm^2/sec by Dubin et al. [27].

The third virus on which electro-optic measurements have been made is fl bacteriophage. This is a circular single stranded DNA phage with flexible filamentous shape. It is a male-specific phage of Hfr E. coli bacteria. It attaches to the F-pili of the bacterium and presumably uses this organelle for injection of its nucleic acid into the cell. Its DNA has a molecular weight of 1.4×10^6, and the total weight of the virus is approximately 11×10^6 [14]. The virus protein is composed of at least two components. The predominant component is the B-protein [32]. A minor component, the A-protein, controls the structural stability of the fl phage to alkaline denaturation [32-34]. Again, amber mutants of this phage, R4 and R5, which contain mutations in gene 3 and gene 6 (which control the amino acid sequence of protein A) are used to analyze the method by which the A-protein stabilizes the structure of the whole phage. The technique used for these

studies by Rossomando and Milstein [12], is steady state dichroism
and steady state dichroism versus field strength. It is similar
to the related virus fd bacteriophage, which has a particle
weight of 11.3×10^6 daltons $(S_{20} = 40)$,   is filamentous in shape
with lengths of 8000 Å × 50 Å diameter. The DNA component
(single-stranded, cyclic) is 1.4×10^6 daltons or about 12.2% w/w.
fd virus apparently has a major part of its protein in an   α
helical form with helix axis at a small angle to the virus axis.
The DNA in the virus has an upstrand and a downstrand with the two
strands twisting about each other [35]. It was of interest to
find the possible orientation of the DNA nucleic acid bases in the
virus by studying the electro-optical behavior of the virus [12].

B.   Internal Structure

To the molecular biologist or virologist the optical
properties of a virus or nucleoprotein are more interesting than
the electrical properties. The optical properties can lead to
information of the orientation and distribution of chromophores in
the particles which can then be associated with a given structural
organization.

In particular the study of the optical behavior of the
nucleic acid bases in the region of absorbance between 200 and
300 nm is emphasized in efforts to discover the organization of
RNA in TMV and DNA in T2, T4, and f1 bacteriophages.

1.   Tobacco Mosaic Virus (TMV)

At the time of writing this review, no one has been
successful in elucidating the in situ orientation of the bases in
the RNA of TMV by x-ray diffraction techniques. It has remained
for optical techniques to obtain information of this orientation
of the optical RNA chromophores and associate this with an actual
physical orientation of the RNA bases.

The internal RNA is approximately 23% hyperchromic and in situ ORD spectra are similar to the ORD pattern of denatured RNA preparations in distilled water or in the presence of urea. This is interpreted as lack of base to base interactions in the virus, and evidence for a more involved interaction between the bases and certain aromatic amino acids by formation of stacking interactions between these groups [36,37].

Franklin [38] interpreted the higher positive birefringence of oriented gels of TMV versus gels of protein as indication that the base planes were parallel to the particle axis. Ultraviolet dichroism of oriented gels of TMV have been studied with similar results [39-41]. It was found that the dichroism was positive between 240 and 295 nm. They suggested that the dichroism at 290 nm was due to a preferred orientation of tryptophan residues. Later, Beaven et al. [42] suggested that dichroism about 260 nm reflected preferred orientation of nucleic acid bases. Flow dichroism measurements by Schachter et al. [43] of the intact TMV and RNA free repolymerized protein rod indicated that the dichroism of RNA was positive at all wavelengths. This indicates that the RNA bases are "approximately" parallel to the long axis of the virus. The positive dichroism of the protein showed that the three tryptophan residues of the protein subunit are "approximately" parallel to the virus axis. The maxima in the dichroic ratio of the oriented protein agree with the maxima and fine structure peak of the absorption spectra of tryptophan [43]. Problems concerning this influence of anisotropic scattering and form dichroism were analyzed and found to be negligible in the interpretation of the data [44,45].

2. T2 and T4 Bacteriophages

The DNA of these viruses, approximately 52 μm long, is pushed into a highly compact space approximately 1000 Å × 850 Å. Thus, the DNA molecule is quite distorted and it is not surprising that it shows many alterations of its physicochemical properties

as compared to in vitro properties. In vivo T2 and T4 DNA are
hyperchromic by about 10 to 13% at 260 nm [46,47]. The ORD
(optical rotatory dispersion), CD (circular dichroism) and MCD
(magnetic circular dichroism) spectra of the DNA internally
packaged is very different from the released DNA of the disrupted
virus [24-26,46]. Tikchonenko et al. [47] interpret the
hypochromicity as evidence for single-stranded regions in the
internal DNA package. These regions may be interacting with the
interval proteins to help stabilize the condensed DNA package.
On the basis of studies on CD films of DNA on the A, B, and C
geometries, Tunis-Schneider and Maestre [48], Nelson and Johnson
[49], and Dorman and Maestre [46] have suggested that the DNA inside T2, T4, and T6 phages are in the C geometry which is the most
compact form ever measured by x-ray analysis of DNA fibers [50].
The orientation of T2 and T4 DNA has been studied by x-ray
scattering on films and fibers oriented virus by North and Rich
[5], and Maestre and Kilkson [51]. Bendet et al. [52] have measured the birefringence pattern of films of bacteriophage.
Qualitiatively, these workers agree that there is a preferred
orientation of the DNA parallel to the main axis of the phage (see
Fig. 1).

Quantitative measurements on the flow birefringence of T4
bacteriophage and T4er 56, a tryptophan T4 requiring mutant and
their ghosts by Gellert and Davies [53] gave results that about 9%
of the DNA was preferentially aligned along the axis of the phage.
The thermal stability of the oriented package of DNA was measured
by UV absorbance, flow birefringence, and sedimentation velocity
by Inners and Bendet [54] showing that the tertiary structure of
the internal DNA remained intact even at temperatures up to $20^{\circ}C$
higher than the melting temperature of free DNA under comparable
conditions.

Electron-microscopic studies by Cummings and Wanko [20]
showed the existence of a DNA free volume inside the head of T2
phage of dimensions 70×150 Å parallel to the axis of the phage.

This DNA free volume was predicted by Cole and Langley [55] from electron inactivation measurements by Kilkson and Maestre [56] from the structure of their proposed model of the packing of DNA as tertiary coiled cols of DNA.

C. Chemical and Structural Stability

The structure of the viruses in consideration vary with respect to their viability, nativity, or structural integrity in the solutions that are used for electro-optic studies, that is of low or zero conductivity. Tobacco mosaic virus and fl bacterio-phage can be safely suspended in deionized triple distilled water without causing any loss of viability and therefore alteration of the structure of the virus. For their experiments in flow dichro-ism Schachter et al. [43] obtained end-to-end aggregation two to three times of the normal TMV length at pH below 5.15 in 0 to 1 M phosphate buffer. Aggregation would be useful if an increase in orientation is desired for certain experiments.

However, double-stranded DNA bacteriophage such as the T-even family and many others of the same morphology cannot tolerate ionic strength much below 0.001 M without suffering inactivation or disruption with concomitant release of the DNA [57]. This condition will limit the applied electric fields needed for the orientation of the particles [58,1]. These types of phages are also very sensitive to pH conditions; the T-even family suffering inactivation at pH below 4.0. Most of the T-phage family can re-main viable in the range from pH 5 to 9 at room temperature in solutions of ionic strengths of 10^{-2} to 10^{-3} M Mg SO4. Mg^{2+} is especially good at maintaining the stability to the phages at the low ionic strength needed for the Kerr effect measurements.

III.   EQUATIONS

The hydrodynamic, optical, and electrical properties of viruses can be obtained by analyzing their electro-optic behavior. The field free relaxation of the signal will give information about the rotary diffusion coefficient. The saturation of birefringence or dichroism as a function of applied field strength will give information as to the optical anisotropy of the virus as the field strength is extrapolated to infinity, i.e.,

$$(g_a - g_b)_{saturation} = \left(\frac{n_\lambda}{2\pi}\right)\left(\frac{\Delta n}{C_v}\right)_{E\to\infty} \qquad \text{[Eq. (35), Chap. 3,}$$

$$\text{with}\quad \Phi = 1]$$

$(\Delta n)_{E\to\infty}$ = birefringence at infinite field strength

A second value for the optical anisotropy is given by the equation of O'Konski et al. [10]

$$(g_a - g_b)_{Kerr\ region} = \frac{n_\lambda}{2\pi}\frac{15}{(b^2 + c)}\frac{1}{C_v}\left(\frac{\Delta n}{E^2}\right)_{E\to 0}$$

which computes the optical anisotropy in terms of the behavior of the particle in the Kerr region as the field approaches zero. In principle these two values should agree unless the particle behaves in a nonlinear fashion with respect to E^2 at very high field strengths.

The analysis of the steady state birefringence or dichroism as a function of the applied field strength will give information on the orienting dipoles (i.e., the electrical properties) of the intact virus or the protein capsid. These orienting dipoles could be mixed in nature, i.e., both intrinsic and induced dipoles. O'Konski et al. [10] have derived expressions for the saturation measurements in the steady state condition. Equation (7) of Chap. 3 may be expressed in the form:

$$\left(\frac{\Delta n/E^2}{\Delta n/E^2}\right)_{E \to 0} = \frac{\Phi(\beta,\gamma)}{\beta^2 + 2\gamma} = \frac{(\Delta\varepsilon/E^2)}{(\Delta\varepsilon/E^2)}_{E \to 0}$$

where $\Phi(\beta,\gamma)$ is an orientation function

$$\beta = (\mu B_1/kT)E = bE$$

$$\gamma = (\alpha_a - \alpha_b)/[kT]E^2 = cE^2$$

$$b = \mu B_1/kT, \quad c = (\alpha_a - \alpha_b)/kT$$

B_1 is the internal field factor   (=1,   usually)

$\alpha_a - \alpha_b$ = anisotropy of the electric polarizability tensor

A plot of $\Delta n/E^2$ or $\Delta\varepsilon/E^2$ as a function of the square of the electric field intensity allows one to obtain the quantities $(\Delta n/E^2)E \to 0$ and $(\Delta\varepsilon/E^2)E \to 0$ which can be used to compute the Kerr constant B

$$B = \frac{1}{(\lambda)} \frac{\Delta n}{(E^2)}_{E \to 0}$$

A convenient method for using the above equations is by the use of the set of master curves computed by O'Konski et al. [10]. A simplification by Yoshioka and O'Konski [59], also utilized by Yamaoka [60], allows the graphical fitting of these curves by graphical superposition of the experimental function on the theoretically computed curve for the best values of $b^2 + 2c$ and b^2 = constant × c. By the use of their relation the anisotropy of the electric polarizability tensor (i.e., induced dipole) and the intrinsic dipole μ of the particles are determined.

Of most interest to the virologist is the optical factor of the virus or nucleoprotein since this can lead to interpretation in terms of a possible model of the structure of the nucleic acid in situ. Both the birefringence and dichroism are amenable to interpretation; the latter being easier to interpret, but more difficult to measure. Superimposed upon the intrinsic birefringence and dichroism and anisotropic light scattering, all of which are size and shape factors of the viruses or nucleoprotein. To get information of the orientation of the chromophores of the nucleic acids and proteins of these particles some method must be found of coping with the form factor perturbations to be able to subtract them experimentally or theoretically from the measured signal.

Form birefringence arises from a scale much larger than molecular, that is when there is an ordered arrangement of optically isotropic material whose size is large when compared with the dimensions of molecules but smaller than the wavelength of light. Only two specific cases have been solved exactly, namely, an array of parallel stacked disks and an array of parallel oriented rods by Wiener [61]. If f_1 = fraction of space occupied by oriented disks or plates and $f_2 = 1 - f_1$ fraction of space occupied by surrounding medium. Wiener obtained for the stacked plates

$$n_\varepsilon^2 - n_0^2 = \frac{-f_1 f_2 (n_1^2 - n_2^2)^2}{f_1 n_2^2 + f_2 n_1^2}$$

where n_ε = index of refraction of extraordinary ray

n_0 = index of refraction of ordinary ray

n_1, n_2 are the indexes of refraction of dielectric disks and surrounding media

For oriented parallel rods the formula obtained by Wiener, if two rods occupy a small fraction of the volume $f_1 \ll 1$, we have

$$n_\varepsilon^2 - n_0^2 = \frac{f_1 f_2 (n_1^2 - n_2^2)^2}{f_1 n_2^2 + f_2 n_1^2}$$

always giving positive birefringence. In both cases if it is possible to vary the index of refraction of the media n_2 the form birefringence will disappear at

$$n_1 = n_2$$

If both forms of birefringence are present a graph of $|n_\varepsilon^2 - n_0^2|$ plotted against n_2 will show a nonzero minimum at $n_2 = n_1$.

Form birefringence can be a considerable factor in the interpretation of particles whose intrinsic birefringence is low and it is always a problem in the interpretation of the electric birefringence in such viruses as TMV virus and T2, T4, and fl bacteriophage. The influence of form dichroism and the electric dichroism of TMV has been analyzed by Mayfield and Bendet [44,45] and was shown to be a negligible factor in this case. For highly scattering particles, anisotropic light scattering is probably of more concern in the interpretation of electric dichroism signals, particularly in the region of the spectrum where the viruses have low absorption. Approximate corrections have been made by extrapolating the anisotropic scattering in the nonabsorbing regions of the spectrum, using an inverse fourth power low on the wavelength (Raleigh scattering behavior) [62]. For an enlightening discussion and detailed application of the form birefringence problem, see Bendet and Bearden papers on the "Birefringence of Spermatozoa" [31].

IV. ELECTRO-OPTIC MEASUREMENTS

A. Tobacco Mosaic Virus

TMV is the most studied nucleoprotein or virus in the history of electric birefringence (an extensive list of references can be found in the review by Stoylov [131]). Lauffer was the first to report birefringence of TMV in sinusoidal varying electric fields [63]. Since then, the electro-optic studies on the orienting mechanism for TMV in transient electric fields are classic [6-11, 64]. All strains of TMV have largely induced dipole orienting

mechanisms with the exception of the "Holmes Rib Grass" strain
which has a considerable intrinsic dipole (1.0×10^{-13} esu)
parallel to the viral main axis [9].

According to Lauffer all of the birefringence, which is
positive, is due to the form birefringence factor [63]. It is
obvious that such a rod shaped object should have a very strong
form birefringence according to Wiener's equations [61].

Electric dichroism studies by Allen and Van Holde [62] on
TMV shows the behavior of the saturation dichroism $(\Delta\varepsilon/\varepsilon)_{sat}$ as
a function of pH. In their conclusion Allen and Van Holde [62]
agree with O'Konski and Haltner [7,8] in assigning the orientation
of particles of the common TMV strain to a pure induced dipole
mechanism. Of more interest to the question of virus structure is
Allen and Van Holde's studies of the saturation dichroism as a
function of wavelengths [62]. To interpret this data a correction
due to anisotropic scattering by using a fourth power law extrapolation was performed by the following formulas

$$\frac{\Delta\varepsilon}{\varepsilon} = \left(\frac{\Delta\varepsilon}{\varepsilon}\right)_{sc}\left(\frac{\varepsilon_{sc}}{\varepsilon}\right) + \left(\frac{\Delta\varepsilon}{\varepsilon}\right)_{a}\left(\frac{\varepsilon_{a}}{\varepsilon}\right)$$

where $\varepsilon = \varepsilon_{\parallel} - \varepsilon_{\perp}$
        a = absorbance
        sc = scattering

The quantity $(\Delta\varepsilon/\varepsilon)_{sc}$ is then the apparent dichroism due to
scattering. Figure 2 shows the plot of the corrected and uncor-
rected dichroism as a function of wavelength [62]. These authors
concluded that there are some $\pi-\pi^*$ transitions due to the
aromatic amino acids centered near 280 nm and a transition of
opposite orientation near 300 nm (probably $n-\pi^*$ in nature).
They concluded from the modest magnitude of the dichroism that
there is some preferential orientation of the aromatic amino
acids and nucleotide residues with rings parallel to the helix
axis. Allen and Van Holde's data [62] agree very well with the
flow orientation data of Mayfield and Bendet [45], with

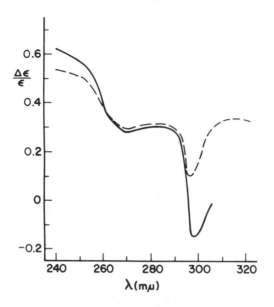

FIG. 2. Variation of TMV dichroism (electro-optic orientation) at pH 7.5 with wavelength: (---) before and (———) after correction for scattering. (From Allen and Van Holde [62].)

substantially similar conclusions.

B. T2, T4, T5, and T7 Bacteriophages

T2 intact bacteriophage and T2 ghost were first measured by electric birefringence by Maestre [58] (Fig. 3). He determined the specific birefringence, $\Delta n/c$, the intrinsic anisotropy of the electric polarization and the rotational diffusion coefficients of both of the intact T2$^+$ (lysis inhibition strain) phage and its ghost. The field free relaxation measurements show both the intact virus and its ghost to have two rotary diffusion coefficients. These coefficients have values of 555 ± 54 and 111 ± 22 sec^{-1} for the phage. In order to interpret these coefficients, Filson and Bloomfield [65] using their multicomponent hydrodynamic theory calculated these parameters for various

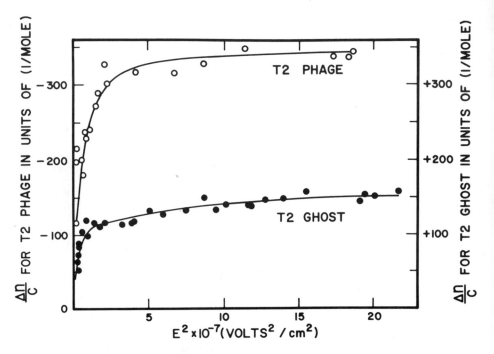

FIG. 3. Birefringence saturation curves for T2 virus and
ghost: (O) curve for T2 virus refers to left-hand scale; (●)
curve for T2 ghost refers to the right-hand scale. (From Maestre
[58].)

conformational models of the phage. Later Douthart and Bloomfield
[66] did experiments with macroscopic models. For the model with
no extended fibers and long head geometry (1280 Å) they obtain
from theory, 552 sec^{-1}, from macroscopic model 556 sec^{-1} in very
good agreement with the electro-optic value. For extended fibers
straight or kinked at an angle of 3/4 π radians from the axis of
the tail they got 130 and 139 sec^{-1}, again in reasonable agreement with the long relaxation component. The ghost had rotary
diffusion coefficients of 688 ± 89 and 1 and 161 ± 29 sec^{-1}. The
saturation of the specific birefringence of the free nucleic acid
parallel to the main axis of the phage in close agreement with
Gellert and Davies [53]. The analysis of birefringence versus
applied field strength in the Kerr region (Fig. 4) gave the

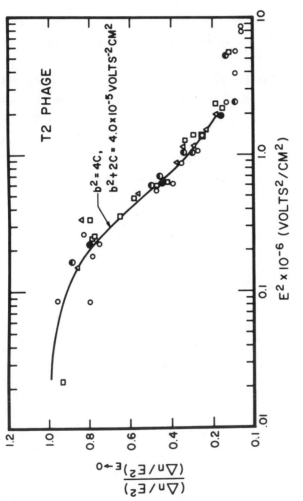
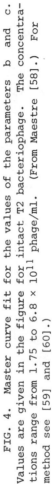

FIG. 4. Master curve fit for the values of the parameters b and c. Values are given in the figure for intact T2 bacteriophage. The concentrations range from 1.75 to 6.8 × 10^11 phage/ml. (From Maestre [58].) For method see [59] and [60].)

values for the anisotropy of the polarizability $\alpha_a - \alpha_b$ and
intrinsic dipole of both phage and ghost: for T2 phage
$\alpha_a - \alpha_b = 5.0 \times 10^{-14}$ cm^3 and $\mu = 64,400$ D; for T2 ghost
$\alpha_a - \alpha_b = 7.9 \times 10^{-14}$ cm^3 and $\mu = 57,200$ D. The high intrinsic
dipoles for phage and ghost are interpreted to be associated with
the coat protein, specifically with the tail spikes or tail fibers
and having some function in the mechanism of attachment of phage
to host cell wall [17].

Recently similar experiments by Greve [1] were done on T4
bacteriophage and an amber mutant of T4 named X4E that when grown
on the restrictive host (E. coli B) produces an intact X4E particle that lacks all six of the tail fibers. Greve [1] found that
the Kerr coefficient, rotational diffusion coefficient and
sedimentation coefficient of the T4B (with tryptophan added to
produce extension of the tail fibers) were dependent on buffer
concentration and reached plateau values at high concentration
given by (Greves units cm^2 OD^{-1} (statvolt^{-2}) converted to esu
units for comparison)

$$K_{sp} = 02.52 \pm 0.17 \times 10^{-3} \text{ cm}^5 \text{ g}^{-1} \text{ statvolts}^{-2}$$

$$\theta_{25,w} = 133 \pm 4 \text{ sec}^{-1} \text{ and } S_{20,w} = 818 \pm 11 \text{ S}$$

The values for the virus in the nonabsorbing state (fiber con-
tracted) are:

$$\theta_{25,w} = 280 \pm 9 \text{ sec}^{-1}, \quad S_{20,w} = 1023 \pm 12 \text{ S}$$

and the Kerr coefficient had values

$$K_{sp} = K_{sp} = (-1.29 \pm 0.4 \times 10^{-5} \text{ to } -4.22 \pm 0.3)$$
$$\times 10^{-3} \text{ cm}^5 \text{ g}^{-1} \text{ statvolt}^{-2}$$

Greve concludes from his data that the extension of the tail fiber structure accounts for the variation in rotational diffusion coefficient and sedimentation velocity of T4 virus. Greve also showed by the reverse pulse transient technique in agreement with Maestre [58] that this bacteriophage has a large intrinsic dipole associated with its extension of the tail fiber structure. The values for the saturation birefringence obtained by Greve range from $\Delta n_s/C = 275$ to -370 mole^{-1} (at 546 nm). If the value of the form birefringence for the ghost $(\Delta n_s/C)_{ghost} = +133$ is subtracted from the above figures we would have a $(\Delta n_s/C)_{internal\ DNA} = -408$ to -503 mole^{-1}. This when divided by the specific birefringence of DNA in solution of -1800 and in fibers of -3200 mole^{-1} would give an orientation factor for the internal DNA of as low as 12.6% to a maximum as high as 28% of the DNA is aligned parallel to the main axis of phage.

Recently T4, T5, T7 bacteriophages were studied by Kwan et al. [67]. The rotary diffusion coefficients in 10^{-10} m Tris-HCl, 10^{-4} Mg SO$_4$ buffer, pH 6.7 for T4 was 329 ± 8 sec^{-1}, for T5 was 1157 ± 150 sec^{-1}, and T7 was 2825 ± 114 in 10^{-3} Mg SO$_4$ pH 7.3. No saturation values for orientation were obtained for T5 and T4 phages. The saturation specific birefringence for T4 bacteriophage was found to be $\Delta m_s/c = -202 \pm 20$ (liters/mol), which agrees quite well with the T2 phage value [58]. The specific Kerr constants where for T4, -4.49×10^{-9}, for T5, -8.90×10^{-10}, and for T7, $+4.57 \times 10^{-10}$. The optical anisotropy factor for T4 was found to be -3.2×10^{-4} in the Kerr region, -2.9×10^{-4} in the saturation region, again very similar to those obtained for T2 [58]. The analysis of the optical properties enable these authors to reject a ball model for the packing of the DNA in bacteriophage heads in favor of the spool model for DNA packing by Richards et al. [68]. The mean orientation of DNA was interpreted by Kwan et al. [67] to be oriented parallel to main phage axis in T4, T5 bacteriophage and perpendicular to main phage axis in T7 bacteriophage.

C. fl Bacteriophage

The other type of bacteriophage in which electro-optic measurements have been made is fl bacteriophage [3,12]. Both electric birefringence and electric dichroism measurements were done on the wild type fl virus and on amber mutants R4 which grow on Ser-I host (amber triplet translated into serine) on Ser-II (amber triplet into glutamine) and Ser-III (amber triplet into tyrosine) in the A-protein. All four particles completely oriented by 12,000 V/cm applied field strength. The differences in saturating values of the birefringence were interpreted as differences on the optical anisotropy due to the arrangement of the nucleic acid and protein in the virus. The dichroism at 260 nm due to the DNA is (-0.3) approximately the same for fl (wild type) and R4 (Ser), indicating that the DNA base planes of these two phages have the same orientation. However, at 280 nm $\Delta\varepsilon$ is +0.5 for fl (wild type) and R4 (tyrosine) but smaller in value for R4 (serine) reflecting a change for the protein chromophores for this virus. In the Kerr region fl and R4 (Ser) follow the Kerr law but R4 (glutamine) and R4 (tyrosine) deviate from linearity. The authors interpreted this deviation as a sign of extra structural flexibility [69] for these two viruses indicating that there are two structurally different forms of the virus mediated by the gene 3 product. Milstein and Rossomando continued their electro-optic studies of fl and R4 by altering the structure with heat treatment of both viruses [12]. At various temperatures they found a change in the packing arrangement of the B-protein subunits causing the R4 to behave as a particle having similar hydrodynamic properties to the fl. The kinetics of the heat effect on the electro-optic measurements indicated that R4 is stabilized relative to fl at higher temperatures. There is also an increase in the tilt of the DNA bases of 1.5° as the result of heating. Milstein and Rossomando [3,12] presented models to account for these changes in viral structure.

The fd virus, which is very similar in structure to fl virus, was studied by flow dichroism by Bendet and Mayfield [70]. They found the dichroism to be positive for wavelengths larger than 262 nm and shorter than 239 nm while being negative in the 239 to 262 nm region. They applied a simple theory to arrive at orientation parameters for the DNA base planes inside the virus. The normals to the DNA base planes have a tilt of 20° to 30° with respect to the fd virus axis. The amino acid residues responsible for the 280 nm absorption are oriented on the average more parallel than perpendicular to the long axis of the virus. They also found that form dichroism did not make an important contribution to the dichroism of the virus [44,45].

D. Nucleohistones

The electro-optic properties of the soluble fraction of deoxyribonucleohistone (S-DNH) from calf thymus was first studied by Golub and Dvorkin [71]. They reported that in a field of 1 to 2 kV/cm: (a) the negative sign of the birefringence indicates an orientation of the base wings of the DNA perpendicular to the major axis of the DNH molecules; (b) the relaxation times are consistent with those of dimers, found by end to end joining of the monomer molecules; and (c) the values of the optical retardation were one-half that obtained from DNA (due to a smaller anisotropy of the optical polarization of the DNH) at the same ionic strength and applied field.

The electro-optic behavior of the gel fraction (i.e., large molecular weight chromatin) of deoxyribonucleohistone (gel-DNH) were studied by Houssier and Fredericq [72,73]. These authors reported that gel-DNH would show electro-optic saturation of high field strength (14 kV/cm) and that the field free relaxation times increased sharply with decreasing field strength. These were interpreted as the effect of histones forming somewhat

regular aggregates with a wide distribution of sizes. The
decrease in birefringence with increasing ionic strength was
interpreted by Houssier and Fredericq as due to the disordering
of the aggregates. The mean values obtained for the relaxation
times were τ_{min} = 80 to 90 µsec, τ_{max} = 3 to 4 msec at 13
to 14 kV/cm (10 to 10 mg/dl concentration, H_2O and pH 6.5). In
1 mM NaCl pH 6.5, they obtained τ_{min} = 45 to 50 µsec and
τ_{max} = 0.6 to 0.7 msec. These results are consistent with the
presence of aggregates in the gel-like solutions. When the his-
tone was removed by enzymic degradation, the optical anisotropy
increased up to values corresponding to those of the DNA. The
spectra of the UV dichroism indicated the presence of an n-π*
transition approximately at 280 to 290 nm. Other transitions were
also detected below 260 nm which were not perpendicular to the
helix axis.

Houssier and Fredericq [73] then proceeded to study the
electro-optic properties of DNA-proflavine complexes. They found
that this complex displayed a negative dichroism in the visible
range due entirely to the dye molecule. The dichroic ratio D
increased with decreasing number of dye molecules bound per atom
of phosphorous (r = ratio) and reached a maximum value at about
1.85 (for $r \leq 0.1$), approximately constant in the whole visible
absorption band (at 13 to 13.5 kV/cm). The birefringence relaxation times were not noticeably affected by the interaction except
for $r \geq 0.10$ where a decrease in τ was observed. These
authors interpreted their result as an intercalation of the proflavine cation between adjacent nucleotide pairs, for at most one
proflavine molecule per five nucleotide pairs in the gel DNH.

E. Ribosomes

Similar electro-optic proflavine binding studies were made by
Schoentjes and Fredericq [74-76] on native and "unfolded ribosomes." They showed that the intercalation binding sites of the
dyes involve two subtypes of sites. These sites can be related to
different nucleotide compositions and secondary structure of ribosomal ribonucleic acid (rRNA) regions. These authors interpreted
the negative dichroism of the bound dye at 458 nm when bound to
the rRNA within the yeast ribosome to agree with the model proposed by Cox and Bonanou [77] in which the base planes of the
nucleotides are not parallel in the native ribosome but rather
radiate around the folding axis of the ribonucleoprotein sheet
[76].

V.   CONCLUSIONS

This review shows that a considerable amount of information
can be obtained about the internal arrangement of large biological
structures by the use of electro-optic techniques. This technique
would be more useful and more popular if it could be automated in
the same fashion as spectrophotometry. There are a large number
of viruses and nucleohistone complexes on which it would be very
profitable scientifically to have comparative measurements of the
electro-optic behavior. Particularly the optical anisotropy tensor as a function of wavelength in the 200 nm to 350 nm region
would yield a rich harvest of information on the very important
π-π* transitions of DNA and RNA when complexed with proteins and

histones in biologically active particles like viruses, ribosomes, and chromosomes.

REFERENCES

1. J. Greve, Thesis, University of Amsterdam, 1972.

2. M. F. Maestre, Am. Chem. Soc. Polymer Reports, 7, 1163 (1966).

3. J. B. Milstein and E. F. Rossomando, Virology, 46, 655 (1971).

4. A. Norman and J. A. Field, Arch. Biochem. Biophys., 70, 257 (1957).

5. A. C. T. North and A. Rich, Nature, 191, 1242 (1961).

6. C. T. O'Konski and B. H. Zimm, Science, 111, 113 (1950).

7. C. T. O'Konski and A. J. Haltner, J. Am. Chem. Soc., 78, 3604 (1956).

8. C. T. O'Konski and A. J. Haltner, Am. Chem. Soc., 79, 5634 (1957).

9. C. T. O'Konski,and R. M. Pytkowicz, J. Am. Chem. Soc., 79, 4815 (1957).

10. C. T. O'Konski, K. Yoshioka, and W. H. Orttung, J. Phys. Chem., 63, 1558 (1959).

11. C. T. O'Konski and S. J. Krause, J. Phys. Chem., 74, 3243 (1970).

12. E. F. Rossomando and J. B. Milstein, J. Mol. Biol., 58, 187 (1971).

13. S. P. Stoylov, Advan. Colloid Interface Sci., 45, 110 (1971).

14. H. Fraenkel-Conrat, The Chemistry and Biology of Viruses, Academic, New York, 1969.

15. D. L. D. Caspar, Adv. Protein Chem., 18, 37 (1963).

16. H. Fraenkel-Conrat and R. C. Williams, Proc. Nat. Acad. Sci. U.S., 41, 690 (1955).

17. G. Stent, Molecular Biology of Bacterial Viruses, Freeman, San Francisco, Calif., 1963.

18. D. J. Cummings and L. M. Kozloff, J. Mol. Biol., 5, 50 (1962).

19. D. J. Cummings and L. M. Kozloff, Biochim. Biophys. Acta, 44, 445 (1960)

20. D. J. Cummings and T. Wanko, J. Mol. Biol., 7, 658 (1963).

21. S. Brenner, G. Streisinger, R. W. Horne, S. P. Champe,
     L. Barnett, S. Benzer, and M. W. Rees, J. Mol. Biol., 1, 281
     (1959).

22. A. E. Hook, A. R. Beard, D. Taylor, G. Sharp, and L. W.
     Beard, J. Biol. Chem., 165, 241 (1946).

23. R. M. Herriott and J. L. Barlow, J. Gen. Physiol., 40, 808
     (1956).

24. M. F. Maestre, D. Gray, and R. Cook, Biopolymers, 10, 2537
     (1971).

25. M. F. Maestre and I. Tinoco, Jr., J. Mol. Biol., 12, 287
     (1965).

26. M. F. Maestre and I. Tinoco, Jr., J. Mol. Biol., 23, 323
     (1967).

27. S. B. Dubin, G. B. Benedeck, F. C. Bancroft, and D. J.
     Freifelder, Mol. Biol., 54, 547 (1970)

28. E. Edgar, Proc. Nat. Acad. Sci. U.S., 55, 498 (1966).

29. S. Brenner and R. W. Horne, Biochem. Biophys. Acta., 34, 103
     (1959).

30. D. Bradley, Bacteriol. Rev., 31, 320 (1967).

31. J. Bearden, Jr., and I. Bendet, J. Cell. Biol., 55, 489-501
     (1972).

32. D. Pratt, H. Tzagoloff, and J. Beavdoin, Virology, 39, 42
     (1969).

33. T. J. Henry and D. Pratt, Proc. Nat. Acad. Sci. U.S., 62, 800
     (1969).

34. E. F. Rossomando and N. D. Zinder, J. Mol. Biol., 36, 387
     (1968).

35. D. A. Marvin, J. Mol. Biol., 15, 8 (1966).

36. E. N. Dobrov, I. A. Andriashivili, and T. I. Tikchonenko, J.
     Gen. Virol., 16, 161 (1972).

37. E. N. Dobrov, L. A. Mazhul, S. V. Kust, and T. I.
     Tikchonenko, Mol. Biol. (Russ.), 7, 254 (1973).

38. R. E. Franklin, Biochim. Biophys. Acta., 18, 313 (1955).

39. A. Butenandt, H. Friedrich-Freksa, S. Hartwig, and
     G. Scheibe, Hoppe-Seyl. 2, 274, 276 (1942).

40. M. F. Perutz, E. M. Jope, and R. Barer, Disc. Faraday Soc.,
     9, 423 (1950).

41. W. E. Seeds and M. H. F. Wilkins, Disc. Faraday Soc., 9, 417
     (1950).

42. G. H. Beaven, E. R. Holiday, and E. A. Johnson, Nucleic Acids, 1, 532 (1955).

43. E. Schachter, I. J. Bendet, and M. A. Lauffer, J. Mol. Biol., 22, 165 (1966).

44. J. E. Mayfield and I. J. Bendet, Biopolymers, 9, 655 (1970).

45. J. E. Mayfield and I. J. Bendet, Biopolymers, 9, 669 (1970).

46. B. Dorman and M. Maestre, Proc. Nat. Acad. Sci. U.S. (1973).

47. T. I. Tikchonenko, E. N. Dobrov, G. A. Velikodvorskaya, and N. P. Kisseleva, J. Mol. Biol., 18, 58 (1966).

48. M. J. B. Tunis-Schneider and M. F. Maestre, J. Mol. Biol., 52, 521 (1970).

49. R. G. Nelson and W. C. Johnson, Jr., Biochem. Biophys. Res. Commun., 41, 211 (1970).

50. D. A. Marvin, M. Spencer, M. H. F. Wilkins, and L. D. Hamilton, J. Mol. Biol., 3, 547 (1961).

51. M. F. Maestre and R. Kilkson, Nature, 193, 366 (1962).

52. I. J. Bendet, D. A. Goldstein, and M. A. Lauffer, Nature, 187, 781 (1960).

53. M. Gellert and D. R. Davies, J. Mol. Biol., 8, 342 (1964).

54. D. Inners and I. J. Bendet, Virology, 38, 269 (1967).

55. A. Cole and R. Langley, Biophys. J., 3, 189 (1963).

56. R. Kilkson and M. F. Maestre, Nature, 195, 494 (1962).

57. T. F. Anderson, Cold Spring Harbor Symp. Quant. Biol., 18, 197 (1953).

58. M. F. Maestre, Biopolymers, 6, 415 (1968).

59. K. Yoshioka and C. T. O'Konski, Biopolymers, 4, 499 (1966).

60. K. Yamaoka, Thesis, University of California, Berkeley, 1964.

61. O. Wiener, Kolloidchem. Beih., 23, 189 (1926).

62. S. Allen and K. E. Van Holde, Biopolymers, 10, 865 (1971).

63. M. A. Lauffer, J. Am. Chem. Soc., 61, 2412 (1939).

64. I. Tinoco, Jr. and K. Yamaoka, J. Phys. Chem., 63, 423 (19).

65. D. P. Filson and V. A. Bloomfield, Biochemistry, 6, 1650 (1967).

66. R. J. Douthart and V. A. Bloomfield, Biochemistry, 7, 3912 (1968).

67. M. Kwan, M. F. Maestre, C. T. O'Konski, and L. S. Shepard, in preparation (1977).

68. K. E. Richards, R. C. Williams, and R. Calendar, J. Mol. Biol., 78, 252 (1973).

69. J. B. Milstein and E. Charney, Macromolecules, 2, 678 (1969).

70. I. J. Bendet and J. E. Mayfield, Biophys. J., 7, 111 (1967).

71. E. I. Golub and G. A. Dvorkin, Dokl. Akad. Nauk USSR, 151, 224 (1963).

72. C. Houssier and E. Fredericq, Biochim. Biophys. Acta., 88, 450 (1964).

73. C. Houssier and E. Fredericq, Biochim. Biophys. Acta., 12, 113 (1966).

74. M. Schoentjes and E. Fredericq, Arch. Int. Physiol. Biochim., 76, 947 (1968).

75. M. Schoentjes and E. Fredericq, Arch. Int. Physiol. Biochim., 78, 175 (1969).

76. M. Schoentjes and E. Fredericq, Biopolymers, 11, 361 (1972).

77. R. A. Cox and S. A. Bonanou, Biochem. J., 114, 769 (1969).

78. R. D. Camerini-Otero, R. M. Franklin, and L. Day, Biochemistry, 13, 3763 (1974).

Chapter 21

ELECTRO-OPTICS OF NERVE MEMBRANES

R. D. Keynes

Physiological Laboratory
University of Cambridge
Cambridge, England

I.   INTRODUCTION

Virtually all living cells are bounded by membranes whose
electrical resistance is several orders of magnitude greater than
that of the cytoplasm or the extracellular medium. When, therefore, a gradient of electric potential is applied across either an
organized tissue or a suspension of separate cells, the cells are
effectively short-circuited by the low resistance shunt pathways
of the extracellular fluids between them. Under these circumstances, there may be optical effects associated with the
electrophoretic migration of any cells that are free to move [1],
especially if the surface charge is asymmetrically distributed, as
in spermatozoa [2]. But since such effects are not directly related to electrically induced changes at the cellular or molecular
levels, they need not be considered here. A situation of greater

interest in the present context arises when the potential
difference between the inside and outside of a cell alters, as it
does during the propagation of impulses in nerve and muscle
fibers. The potential drop (PD) is then confined to the immediate
vicinity of the membrane, because of its relatively high resistance; and although the total pd is only around 100 mV, the electric field across the membrane dielectric may change by as much as
400,000 V/cm, because its thickness is only of the order of 25 Å.
With such large fields involved, it is not surprising that there
are detectable changes in the optical properties of the membrane
which may reflect structural transitions [3] and as explained in
Chap. 3 should provide valuable information about its molecular
organization. As will be seen, there are also other optical
changes in nerve fibers which depend on current flow rather than
membrane potential, but these are in a sense secondary effects of
less importance.

II. THE OPTICAL RETARDATION OF NERVE MEMBRANES

Several of the structural components of a nerve fiber are
birefringent. Some parts have an optic axis parallel to the long
axis of the fiber, but others may have a radial or circumferential
optic axis [4]. It is in general simplest to refer to the sign of
the retardation that is observed as if the axon were uniaxially
birefringent with a longitudinal optic axis. With this conven-
tion, the net retardation of a squid axon or a crab nerve trunk is
positive. The average retardation measured at the center of a
squid axon 850 μm in diameter was 56 nm [4], which represents a
birefringence Δn of 6.6×10^{-5}. The average birefringence of
the much smaller fibers in a crab leg nerve is about 5×10^{-5} [4].
From the earlier studies of Schmitt and Bear [5,6] it seems clear
that most of the resting retardation at the center of the axon
arises from the presence of longitudinally oriented fibrils and

tubules in the axoplasm; removal of the axoplasm reduces the
retardation at the center of a squid axon by 90% [4]. At the
edges of the axon, the positive retardation is contributed mainly
by the form and intrinsic birefringence of the Schwann cell and
connective tissue [5,6]. The axon membrane itself contributes
relatively little to the net resting birefringence. If, as seems
very likely, it has a structure similar to that of the erythrocyte
ghosts studied by Mitchison [7], it would have a positive intrinsic birefringence and a negative form birefringence both with a
radial optic axis, so that referred to the longitudinal axis its
intrinsic birefringence would make a negative contribution and its
form birefringence a positive one. However, the retardation of
the edge of an erythrocyte ghost, placed in glycerol to minimize
the contribution of form birefringence, is only about 0.4 nm [7],
or less than 1% of the retardation of a squid axon. Since the
nerve membrane is thus a rather minor component as far as the
overall birefringence is concerned, the absolute magnitude of its
retardation has not yet been determined.

In collaboration with R. S. Bear, Schmitt and Schmitt [8]
used a photocell technique to look for changes in the net birefringence of a squid axon during impulse propagation, but had to
conclude that the retardation did not alter by more than 0.1%.
Repeating the experiment 30 years later, we [4,9,10] were more
successful, thanks to the availability of a signal averaging computer which made it possible to detect intensity changes as small
as 1 part in a million. Figure 1 shows simultaneous recordings of
the retardation change and the action potential in a squid giant
axon. One reason for locating the optical change in the membrane
is that the optical signal was the same whether an intact axon was
used, or whether, as on this occasion, the axoplasm was replaced
by a potassium fluoride solution. Another check on the localiza-
tion was to measure the signal at different positions across the
axon selected by placing a narrow slit at the image plane of the
polarizing microscope. The results agreed perfectly with the

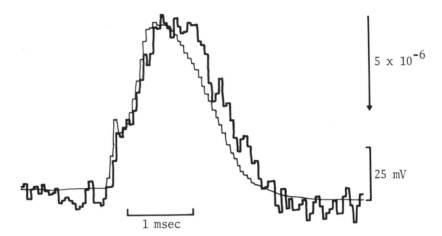

FIG. 1. Simultaneous recordings of the optical retardation change (thick line, average of 2030 sweeps) and the action potential in a squid giant axon perfused with a potassium fluoride solution. Temperature 14°C. The direction of the arrow to the right of the tracings indicates an increase in light intensity, and its length corresponds to the stated value of $\Delta I/I_0$ for a single sweep. (From Ref. 4.)

predicted retardation for a thin birefringent cylinder with a radial optic axis. Neither of these observations establishes conclusively that the retardation change takes place in the membrane rather than in the Schwann cell that is wrapped around the axon, but the absence of any potential drop across the Schwann cell comparable with that across the membrane makes the possibility a remote one. Moreover, precisely similar changes in optical retardation are observed in the electroplates of the electric eel [11], where there are no Schwann cells to complicate the issue.

All the obvious controls have been performed to verify that the signal illustrated in Fig. 1 arises wholly from a change in retardation, and is not measurably affected by changes in light scattering, optical rotation, or dichroism. Probably the most convincing test is the demonstration that subtraction of an appropriate fixed retardation by placing a compensator in series with the axon results in a mirror-image inversion of the intensity change; and there is good quantitative agreement between the added

retardation and the size of the optical signal (see Fig. 6 in Ref.
10).

The optical changes observed during propagation of a nervous
impulse might be caused by structural changes related to
(a) membrane potential changes, (b) the flow of electric current
in local circuits, (c) membrane conductance changes. In order to
distinguish between these possibilities it is necessary to resort
to voltage-clamp techniques [12]. Analysis of the retardation
response in the squid giant axon has shown [10] that it is almost
exclusively a voltage-dependent phenomenon; there may under some
conditions be a small current-dependent component, but no
conductance-linked optical effects have yet been recorded. Detailed examination of the responses observed when hyperpolarizing
voltage-clamp pulses are applied (see Fig. 2) reveals that there
are three components which can be distinguished on a number of
grounds, notably their widely different relaxation times. In a
freshly dissected axon in the condition that we have termed state
1, the response consists of a fast component with a relaxation
time of around 40 μsec at 13°C which gives an increase in light
intensity for an increase in membrane potential, plus a slow
component in the opposite direction with a relaxation time of
about 20 msec (Fig. 2A). Treatment of the axon in various ways
such as exposure to tetrodotoxin, raising the external calcium
concentration, or applying lanthanum, results in an irreversible
transition to state 2, in which the fast component is slowed by a
factor of 5, and the slow reversed component disappears, to be
replaced by one with the same sign as the fast phase but a relaxation time of about 2 msec (Fig. 2D).

In speculating about the molecular origin of the retardation
response it has to be remembered that the quantity observed is the
product of membrane birefringence and membrane thickness. The
compressional force exerted on the membrane by the positive and
negative charge on either side is not inconsiderable, and the
fractional increase in membrane thickness on abolition of a 50 mV
potential has been estimated [10], in agreement with the data of

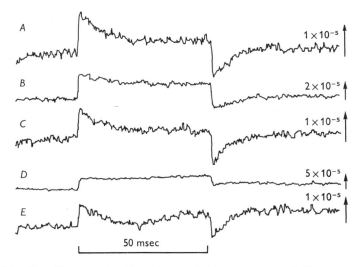

FIG. 2. Changes in the size and time course of the
retardation response observed in a squid giant axon on application
of a 50 mV hyperpolarizing voltage-clamp pulse whose duration was
120 msec in A, C, and D, and 140 msec in B. (A) soon after mount-
ing the axon for observation. (B, C, D) 80, 266, and 540 min
later. Direction and length of arrows as in Fig. 1. 140-338
sweeps averaged. Temperature 14°C. (From Ref. 10.)

Russian workers [13], to lie between 5×10^{-3} and 3×10^{-4}. The
size of the fast component of the retardation response would
correspond to a thickness change of about 8×10^{-4} [10]. Another
reason for identifying the fast component tentatively with an
electrostriction effect is that its sign is correct. A positive
change referred to the longitudinal axis corresponds to a negative
change referred to the radial optic axis of the membrane, so that
raising the potential and further compressing the membrane would
have the observed effect of increasing the net retardation. It
also seems easier to suppose that the molecular reorganization
underlying the transition from state 1 to state 2 involves a
relatively small change in the compressibility of the membrane and
some increase in its relaxation time, than that there is the much
larger alteration that would be implied by an identification with
the component of medium speed seen only in state 2.

The two other components of the retardation response are presumably to be accounted for in terms of a Kerr effect. Unfortunately, it is easier to make this general proposition than to pinpoint the constituents of the membrane that are responsible. A rough calculation indicates [10] that the Kerr constant K_{sp} would need to be of the order of 1×10^{-6}, positive in one case and negative in the other. As may be seen from Table 1 in Chap. 3, there is no difficulty in finding precedents for values of this size, nor would there be if the effect was arising in molecules located at one or other surface of the membrane, and subjected only to a fraction of the total electric field. In unpublished experiments performed in my laboratory by Dr. A. Ikegami, the value of K_{sp} for a typical phospholipid, egg lecithin, measured in a dilute solution in n-decane so as to minimize complications from aggregation phenomena, has been found to be -1×10^{-7}; but the figure might be rather different for molecules arranged in a membrane. The large relaxation times that are observed are harder to explain, though ultimately they should provide a useful clue to the source of each component of the retardation response. There is no reason to think that a polarization or reorientation of more than part of the molecule is involved in each case, and yet even the fastest component has a relaxation time comparable with the time constant for rotation of a whole rhodopsin molecule in a retinal rod membrane [14]. However, allosteric enzymes are known to display relaxation times of 1-20 msec [15]. Structural transitions have been observed in DNA by O'Konski and Stellwagen [3], and the structural changes in membrane macromolecules that underlie the retardation response may be essentially similar in nature.

Another parameter of the retardation response that should have diagnostic value is the relationship between retardation and potential. In the case of the fast component, the results are well fitted by a square law [10], although it is not, as might have been expected, symmetrical about zero membrane potential, but

is displaced well beyond it, indicating the existence of so far
unexplored asymmetries. The size of the component of medium speed,
on the other hand, seems to vary linearly with membrane potential.
For technical reasons, all that can be said about the voltage
dependence of the reversed slow component is that it differs
appreciably from that of the fast component.

Attempts to localize the molecules responsible for the different components of the retardation response by varying the composition of the solution bathing either side of the membrane have
been successful only in the case of the slow reversed component.
However, the size of this component is immediately and reversibly
reduced by raising the external calcium concentration, which suggests that it originates in outward-facing molecules such as the
glycoproteins anchored in the membrane surface.

III.   CHANGES IN LIGHT SCATTERING DURING NERVOUS ACTIVITY

The first observation of an optically detectable change in a
stimulated nerve was made by Hill and Keynes in 1949 [16], studying the light scattered by crab nerve trunks during and after a
train of several hundred stimuli. The effect was then examined in
greater detail by Hill [17], and his findings were later confirmed
and extended to other nerves by Tobias and his colleagues [18-21].
These light scattering changes developed relatively slowly during
the period of activity, and could not be measured satisfactorily
after the passage of single impulses, so that little could be done
to decide on their precise origin. However, the advent of signal
averaging techniques made it much easier to study rapid changes in
light intensity, and when the problem was taken up again in 1967
we soon found [9,22] that there were in fact two components of the
scattering change, the first a transient increase in scattering
that occurs at just about the same time as the action potential,
and the second a persistent decrease that occurs after it and

lasts for some seconds. It is evidently the long-lasting decrease in scattering that produces the cumulative effects investigated earlier. As suggested by Hill [23] the slow component appears to be related to a swelling of the nerve fibers caused by water movements.

Although any signals arising from optical changes located in the membrane are bound to be easiest to detect in nerve trunks containing large numbers of very small fibers, it is again essential to study them by the voltage-clamp technique in isolated axons if much is to be learned about their origin at the molecular level. This means that squid giant axons have to be employed; but fortunately the disadvantage imposed by their relatively small ratio of surface to volume is offset by their ability to survive the application of many thousands of voltage-clamp pulses at a high repetition rate. Thus with the help of a signal averaging computer it was possible to measure light intensity changes down to 1 part in 10^6, and so to analyze the light scattering effects in some detail [24,25]. It turned out that there were several voltage-dependent [24] and several current-dependent [25] components, but again none of the light scattering changes could be directly linked with the mechanisms involved in the alteration of membrane conductance. Both for the action potential and for voltage-clamp steps, the scattering changes were different at right angles (60-120°) and at forward angles (5-30°).

At right angles, a hyperpolarizing voltage-clamp pulse gave a decrease in scattering whose time course could be described by a single exponential with a time constant that averaged 24 μsec at 13°C. At forward angles, the change was in the opposite direction, but had two components whose time constants were 23 and 900 μsec. Since these time constants were about the same size as those of the two fastest components of the retardation response, it is tempting to propose that the 24 μsec component again represents the membrane thickness change, and that the slow component is another manifestation of the Kerr effect seen in state 2 of the

membrane (see p. 749). The voltage-dependence of the two light
scattering changes seemed to be more or less consistent with this
idea [24], but our other findings were harder to reconcile with
it. Thus in an axon in which light scattering and retardation
responses were observed alternately during the transition from
state 1 to state 2, the right-angle scattering never exhibited a
slow component and the forward angle scattering always had one,
while the retardation initially had only the fast component and
finally both of them. Treatment of the axons with butanol or
octanol increased the right angle scattering change significantly,
but failed to affect either forward angle scattering or birefringence. Furthermore, the effect of temperature on the relaxation
time for light scattering seemed to be appreciably smaller $(Q_{10}$
around 1/2) than it was for the retardation response $(Q_{10}$ about
1/8 in state 1) [10,24]. Without knowing more about the precise
nature of the molecular reorientations underlying all these
effects, it is difficult to tell what weight to attach to the
various discrepancies. But it does not seem impossible to envis-
age molecular events that would be revealed as a retardation
change unaccompanied by an alteration in light scattering in one
plane, and vice versa, or whose several optical consequences
would not all be affected to exactly the same extent by modifying
the experimental conditions. The most reliable criterion for
deciding which, if any, of the light scattering and retardation
changes have a common origin, would therefore be an accurate
measurement of their relaxation times. However, the data gathered
so far does not permit us to reach any firm conclusions.

The current-dependent light scattering changes were in
general somewhat larger. Measurements at several different
scattering angles (see Fig. 3) suggested that three components
with distinctly different timecourses were involved. At right
angles, all outward currents, whether carried by K^+ or Na^+ ions,
gave a decrease in scattering that occurred mainly after the
current pulse if its duration did not exceed 5 msec. At angles

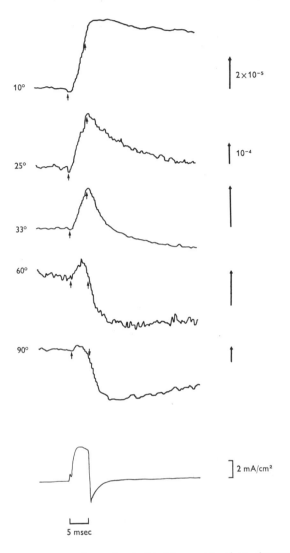

FIG. 3. Current-dependent light-scattering changes given by
a squid giant axon at various angles between the incident light
and the direction of observation. The short arrows near the
scattering traces indicate the beginning and end of the depolarizing voltage-clamp pulse applied to the axon that gave an outward
ionic current whose timecourse is shown below. Direction and
length of arrows to the right as in Fig. 1. 64-256 sweeps averaged. Temperature 13°C. (From Ref. 25.)

around 25°, outward currents gave an increase in scattering whose timecourse followed closely the time integral of the current flow, returning to the baseline with a time constant of 10-40 msec at 12°C. At still smaller angles (10°) the scattering continued to increase for several msec after the end of the current pulse, and returned to the baseline more slowly, with a time constant of about 100 msec. Inward currents gave similar changes in the opposite direction, again with timecourses that were independent of the species of cation carrying the current.

The right angle light scattering change was reversibly reduced in size by about one-third when isethionate ions were substituted for chloride in the sea water bathing the axon, suggesting strongly that a "transport number" effect was involved. Thus if the current crossing the membrane into or out of the restricted space between it and the Schwann cell is carried exclusively by cations, while that flowing onward from the Schwann cell space is carried partly by sodium ions moving in the same direction as the membrane current and partly by chloride ions moving in the opposite direction, there will be a net accumulation or depletion of NaCl in the space, followed by a delayed change in its volume as water moves in or out to restore osmotic balance. In this situation, the sizes of the concentration and subsequent volume changes are directly dependent on the mobility or transport number of the external anion. The timecourse of the right angle scattering change, and the effect of isethionate seawater on its size were therefore consistent with the view that it arose from a shrinking or swelling of the Schwann cell space. The timecourse of the scattering change at 25°, on the other hand, suggested a dependence on the refractive index of the fluid in the Schwann cell space; against this, isethionate seawater had little effect on its size, but it could be argued that there was a counterbalancing alteration in the contribution of the anion to refractive index. As to the origin of the 10° scattering change, no firm conclusion can be reached. In some though not all respects its timecourse could be accounted for by suitably compounding the changes at 25 and 90[0]; but it definitely differed from them in being reversed in sign when the refractive index of the external medium was raised by the addition of dextran or bovine albumin, whereas they were unaffected by this treatment. Little can be said, either, about the slow cumulative changes in light scattering that were the starting point for our recent investigations. We agree with Hill [17,23] in believing that they are associated with a swelling of the axons, but there are difficulties [26] in explaining their magnitude in terms of known movements of sodium, potassium, and chloride ions.

These current-dependent scattering changes only seem to arise because the rate at which ions can diffuse away from the membrane is limited by the presence of the Schwann cell. From the point of view of the neurophysiologist, therefore, they are of some interest in connection with the anatomy of the Schwann cell, but provide no information about the mechanism of excitation. We originally hoped that if -- as still seems very likely to be the case -- there were conformational changes in the molecules constituting the channels through while sodium and potassium ions cross the membrane, they could be detected optically. However, no effects identifiable with the ionic conductance changes on grounds of timecourse or voltage-dependence have yet been seen, either in scattering or in birefringence experiments. Thus in the test illustrated in Fig. 4, a depolarizing voltage-clamp pulse that certainly produced a large sodium conductance change whether sodium current flowed or not, gave the usual scattering change in normal artificial seawater, but no significant optical signal when choline was substituted for sodium. Other methods of preventing current from flowing while still eliciting the conductance increase gave the same result [25]. This negative finding, although unhelpful, is not altogether unexpected. It is now known [27,28] that the sodium channels in nerve membranes are relatively few and far between, there being only some 50 in 1 μm^2. Since in

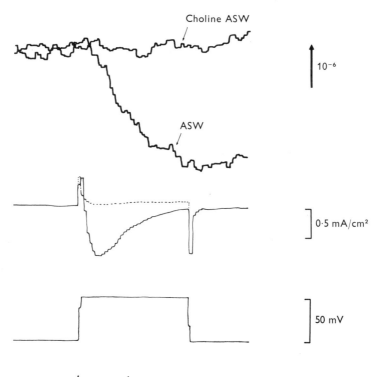

FIG. 4. The changes in forward (25°) light-scattering on
application of a 50 mV depolarizing voltage-clamp pulse (bottom
trace) to a squid giant axon in artificial seawater (ASW) and a
solution in which choline replaced sodium (choline ASW). In ASW,
the depolarization resulted in a large inward current flow
(middle trace, continuous line), which was absent in choline ASW
(middle trace, dashed line). Even though the ionic conductance
increase was similar in the two solutions, the scattering decrease
only occurred when there was a passage of current. 1000 sweeps
averaged in each case. Temperature 13°C. (From Ref. 25.)

the same area there are several million phospholipid and other

molecules, the odds are bound to favor heavily the detection of

any conformational changes in the general structure of the mem-

brane, even if they are small, rather than changes specifically

related to its ionic conductance. Thus it must reluctantly be

concluded that although a detailed examination of the

birefringence and scattering changes should give valuable
information about the behavior of various membrane components
under the influence of an electric field, and thus about the types
of voltage-dependent conformational change that may be expected
to take place in membranes, it is not yet a good approach for
studying the properties of the ionic channels themselves.

IV. FLUORESCENT PROBES IN MEMBRANES

The third optical technique that has recently become
fashionable for membrane studies is that of staining tissues with
a fluorescent dye and examining the changes in fluorescence
emission during electrical activity. The first experiments on
these lines were those of Tasaki et al. [29], who reported that
during the action potential in crab nerves stained with 1-
anilinonaphthalene-9-sulfonate (ANS) there was a fluorescence increase of the order of 1 part in 10^4, whose timecourse was very
similar to that of the electrical signal. Once again, little can
be learnt about the origin of the change at the molecular level
from observations confined to conducted action potentials, and
voltage-clamping has to be applied to make a proper analysis.
The observations of Conti and Tasaki [30] and Davila et al. [31]
on squid axons treated with ANS and a variety of other fluorescent
dyes like rhodamine B, pyronin B, and acridine orange suggest that
the effect is primarily a potential-dependent phenomenon, but the
greatest influence on the fluorescence yield is the dielectric
constant of the immediate environment of the chromophore [32], so
that for those dyes -- the majority -- that are charged, a move-
ment of the molecule into or out of the lipid phase under the
influence of the electric field at the membrane surface is possibly all that is involved. The interaction of ANS with artificial
phospholipid bilayers seems, for example, to be an effect of this
general kind [33]. However, it is then necessary to explain why
some dyes give fluorescence changes dependent on the square of

potential and others on its first power, and why ANS should be
unique in exhibiting optical changes whose signs are determined
by the side of the membrane at which it is added [31]. There are
also intriguing variations in the timecourses of the fluorescence
responses for different dyes [30,31]. Since there is no reason to
suppose that any of the dyes yet tested interact specifically and
exclusively with the ionic channels in nerve membranes, and since
the sparseness of these channels would in any case militate
against obtaining measurable effects from such interactions,
these differences in behavior seem likely to depend on subtleties
of the precise relationship between the structure of individual
dyes and that of the macromolecules at the membrane surface. Thus
fluorescent probes should ultimately yield useful information
about the general structure of cell surfaces. In order to study
the ionic channels themselves, it will be necessary to attach
fluorescent chromophores to specific blocking agents like
tetrodotoxin for sodium or quaternary ammonium compounds for
potassium [34]. But it may prove to be very hard to do this without destroying the high affinity for the binding sites on which
the success of the exercise depends.

Another important application for fluorescent probes is the
determination of changes in membrane potential in cells or
subcellular organelles that are too small for direct measurements
to be made with microelectrodes. This might involve a comparison
of fluorescence yields given by charged and uncharged dyes, as
has been done for mitochondria by Azzi [35]. However, as Radda
[32,36] has explained, the fluorescence yield is affected by
quite a number of factors, and although fluorescent probes
undoubtedly provide a most promising means for studying the
different energy states of mitochondrial membranes, the interpretation of the results in a really critical fashion is still in its
infancy. The same might be said about the application of
fluorescence polarization to examine details of membrane structure
such as its rigidity at different levels and the orientation of
the chromophores of the probing molecules [37].

REFERENCES

1. Cell Electrophoresis, a Symposium convened by the British Biophysical Society (E. J. Ambrose, ed.), Churchill, London, 1965.

2. A. D. Bangham, Proc. R. Soc. B, 155, 292 (1961).

3. C. T. O'Konski and N. C. Stellwagen, Biophys. J., 5, 607 (1965).

4. L. B. Cohen, B. Hille, and R. D. Keynes, J. Physiol. (Lond.), 211, 495 (1970).

5. F. O. Schmitt and R. S. Bear, J. Cell. Comp. Physiol., 9, 261 (1937).

6. F. O. Schmitt and R. S. Bear, Biol. Rev., 14, 27 (1939).

7. J. M. Mitchison, J. Exp. Biol., 30, 397 (1953).

8. F. O. Schmitt and O. H. Schmitt, J. Physiol. (Lond.), 98, 26 (1940).

9. L. B. Cohen, R. D. Keynes, and B. Hille, Nature, 218, 438 (1968).

10. L. B. Cohen, B. Hille, R. D. Keynes, D. Landowne, and E. Rojas, J. Physiol. (Lond.), 218, 205 (1971).

11. L. B. Cohen, B. Hille, and R. D. Keynes, J. Physiol. (Lond.), 203, 489 (1969).

12. Biophysics and Physiology of Excitable Membranes (W. J. Adelman, ed.), Van Nostrand-Reinhold, New York, 1971.

13. G. N. Berestovsky, G. M. Frank, E. A. Liberman, V. Z. Lunevsky, and V. D. Razhin, Biochim. Biophys. Acta, 219, 263 (1970).

14. R. A. Cone, Nature [New Biol.], 236, 39 (1972).

15. G. G. Hammes, Adv. Protein Chem., 23, 1 (1968).

16. D. K. Hill and R. D. Keynes, J. Physiol. (Lond.), 108, 278 (1949).

17. D. K. Hill, J. Physiol. (Lond.), 111, 283 (1950).

18. S. H. Bryant and J. M. Tobias, J. Cell. Comp. Physiol., 40, 199 (1952).

19. S. H. Bryant and J. M. Tobias, J. Cell. Comp. Physiol., 46, 71 (1955).

20. S. N. Shaw and J. M. Tobias, J. Cell. Comp. Physiol., 46, 53 (1955).

21. S. Solomon and J. M. Tobias, J. Cell. Comp. Physiol., $\underline{55}$, 159 (1960).

22. L. B. Cohen and R. D. Keynes, J. Physiol. (Lond.), $\underline{212}$, 259 (1971).

23. D. K. Hill, J. Physiol. (Lond.), $\underline{111}$, 304 (1950).

24. L. B. Cohen, R. D. Keynes, and D. Landowne, J. Physiol. (Lond.), $\underline{224}$, 701 (1972).

25. L. B. Cohen, R. D. Keynes, and D. Landowne, J. Physiol. (Lond.), $\underline{224}$, 727 (1972).

26. P. C. Caldwell and R. D. Keynes, J. Physiol. (Lond.), $\underline{154}$, 177 (1960).

27. R. D. Keynes, J. M. Ritchie, and E. Rojas, J. Physiol. (Lond.), $\underline{213}$, 235 (1971).

28. R. D. Keynes, Nature, $\underline{239}$, 29 (1972).

29. I. Tasaki, A. Watanabe, R. Sandlin, and L. Carnay, Proc. Nat. Acad. Sci. U.S., $\underline{61}$, 883 (1968).

30. F. Conti and I. Tasaki, Science, $\underline{169}$, 1322 (1970).

31. H. V. Davila, L. B. Cohen, and A. S. Waggoner, Biophys. Soc. Abstr., $\underline{11}$, SaAM-E13 (1972).

32. G. K. Radda, Curr. Topics Bioenergetics, $\underline{4}$, 81 (1971).

33. F. Conti and F. Malerba, Biophysik, $\underline{8}$, 326 (1972).

34. C. M. Armstrong, J. Gen. Physiol., $\underline{58}$, 413 (1971).

35. A. Azzi, Biochem. Biophys. Res. Commun., $\underline{37}$, 254 (1969).

36. K. Barrett-Bee and G. K. Radda, Biochim. Biophys. Acta, $\underline{267}$, 211 (1972).

37. I. Tasaki, A. Watanabe, and M. Hallett, J. Membrane Biol., $\underline{8}$, 109 (1972).

Chapter 22

ELECTRO-OPTICS OF LIQUID CRYSTALS

Terry J. Scheffer*
Institut fur Angewandte Feskörperphysik
der Fraunhoder-Gesellschaft
Freiburg, West Germany

Hans Clemens Grüler[†]
Department of Experimental Physics III,
University of Ulm
Ulm, West Germany

---

*Present affiliation: Brown Boveri Research Center, CH-5401
Baden, Switzerland.

[†]Present affiliation: Institut für Angewandte Feskörperphysik
der Fraunhoder-Gesellschaft, Freiburg, West Germany.

I.   INTRODUCTION

In this chapter we shall review many of the electro-optical effects that have been observed in liquid crystals. For lack of space
we shall introduce only those aspects of liquid crystals that are
essential for an understanding of their electro-optical behavior.
We have tried to make this chapter as self-contained as possible, but
the reader is urged to refer to review articles [1-6] and books [7]
dealing with liquid crystals. There are also recent reviews covering
the practical applications of liquid crystals in electro-optical display devices [8].

Liquid crystals represent a state of matter that is intermediate
between the solid crystalline phase and the ordinary isotropic liquid
phase. This intermediate phase is often referred to as a mesophase or
a mesomorphic phase. Liquid crystals flow and adopt the shape of
their containers like ordinary liquids, but they also spontaneously
exhibit anisotropic physical properties ususally only associated with
solid crystals. The anisotropic properties of liquid crystals are the
direct result of the long-range orientational order of the molecules.
The cooperative forces between the molecules in liquid crystals makes
their electro-optical behavior substantially different from that of
solids and liquids. The voltages and fields required to produce appreciable changes in the optical properties in these materials, for
example, are orders of magnitude lower than those usually required for
ordinary liquids. More than just an interaction of the permanent and
induced dipole moments of individual molecules with the electric field
is involved here, because the molecules are nearly completely aligned
in fields where the Langevin equation would predict only one molecule
in 10^4 should be aligned. Even in the isotropic phase near the nematic transition Kerr constants have been measured which are larger than
all known values [9].

We will not discuss lyotropic liquid crystals, because very little is presently known about the electro-optical behavior of this
class of mesophase [10,11]. Lyotropic liquid crystalline phases are

formed when solvents are added to certain substances. There is always
a critical solvent concentration above which the anisotropic phase
transforms to an ordinary isotropic liquid phase. This type of phase
is often formed with aqueous solutions of long-chain saturated fatty
acids [12-14]. Polybenzyl-L-glutamate (PBLG) also forms a lyotropic
system in a variety of solvents [15,16]. The lyotropic liquid crystalline phases play an important role in biological systems [17] and we
believe they will become one of the most interesting subjects in future research.

We shall discuss the more familiar thermotropic liquid
crystals which can be pure substances or mixtures and which adopt
the mesomorphic phase only over a definite temperature interval.
Many thousands of organic compounds form thermotropic liquid crystals when the solid is heated above its melting point. The
liquid crystalline phase appears as a more or less viscous fluid
with a characteristic turbidity, and it can be positively identified by its optical birefringence which is easily detected with a
polarizing microscope. Transitions to other mesophases may occur
in some cases at even higher temperatures, while other compounds
may display only one mesophase. In either case, at another well-

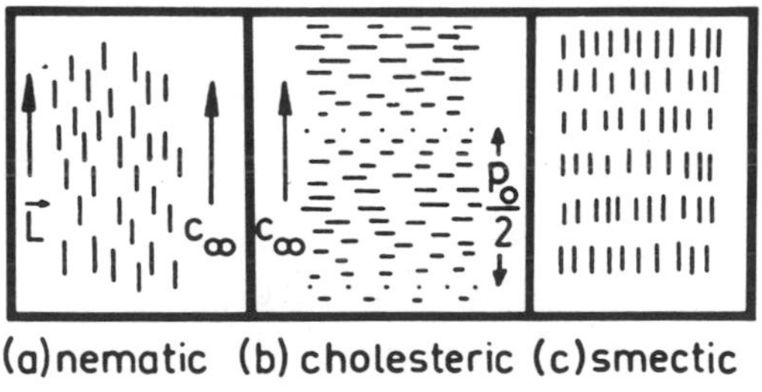

(a) nematic (b) cholesteric (c) smectic

FIG. 1. The three basic thermotropic liquid crystal
structures. The ∞-fold symmetry axis is parallel to the local
optic axis $\underline{L}$ in the nematic phase and perpendicular to it in the
cholesteric phase. P_0 is the characteristic pitch of the
cholesteric structure and is the repeating distance measured along
the helical axis for $\underline{L}$ to twist around by one complete turn.

defined higher temperature, the turbidity of the mesophase suddenly vanishes giving way to the clear appearance of the ordinary liquid. This phase transition is completely reversible and is of first order, having a latent heat of some 100 cal/mole.

Following a proposal by Friedel [18] thermotropic liquid crystals are classified as being either nematic, cholesteric, or smectic. These classifications represent three basic structural types, and they are schematically illustrated in Fig. 1. A consideration of crystal symmetry indicates that a number of other mesomorphic structure types are theoretically possible. Sackmann and Demus [5] and Demus et al. [19] have indeed identified at least five different kinds of liquid crystal mesophases that fit the classical picture of a smectic mesophase. Friedel's simple classification, however, is sufficient for our purposes. We will discuss the structure and resulting physical properties of the nematic, cholesteric and smectic mesophases that are important for an understanding of the electro-optical effects observed in liquid crystals.

A.   Nematic Liquid Crystals

The best known and most often-studied compound exhibiting a nematic liquid crystalline mesophase is p-azoxyanisole (PAA). Many physical constants for PAA have been measured [20]. PAA exists in the nematic state from 118 to 135°C.

$$H_3C\text{-}O\text{-}\langle\bigcirc\rangle\text{-}N\overset{O}{=}N\text{-}\langle\bigcirc\rangle\text{-}O\text{-}CH_3$$

A room temperature nematic liquid crystal compound, 4-methoxybenzylidine-4'-n-butylaniline (MBBA) [21], has also received extensive study. Many of its material constants are known [22]. It has a nematic range from 21 to 48°C.

$$H_9C_4\text{-}\langle\bigcirc\rangle\text{-}N\overset{}{=}\underset{H}{C}\text{-}\langle\bigcirc\rangle\text{-}O\text{-}CH_3$$

A recent review listing other liquid crystalline compounds
that have been used in electro-optical applications has been given
by Castellano [23a]. Gray et al. [23b] have reported on a new
family of nematic liquid crystals having particularly desirable
properties for electro-optical display applications.

1. Spontaneous Orientation and Degree of Order

Molecules that form the nematic mesophase have a rigid,
elongated structure, as is illustrated by PAA and MBBA. The
nematic mesophase is distinguished from an ordinary liquid phase
by a long-range orientational order of the long axes of these
molecules. Averaged over time in a nematic liquid crystal, the
long axes of the molecules in a given microscopic region are all
parallel to a preferred direction which we represent by the unit
vector L. The molecular centers, however, are irregularly
distributed as in an ordinary liquid. This is illustrated in
Fig. 1a. The angular distribution of the molecules about their
long axes is completely random.

It follows that such a structure is uniaxial, the preferred
direction L corresponding to the optic axis. The alignment of
unsymmetrical molecules like MBBA must be nonpolar because effects
like ferroelectricity and the second harmonic generation of light
are not observed. Nehring and Saupe [24] point out that the
existence of certain types of alignment singularities in nematic
liquid crystals is also a direct proof for the absence of a polar
structure.

In reality, thermal energy kT makes the orientation of the
long axis of the individual molecules continuously change direction. The molecules are free to rotate anisotropically and are
on the average, aligned more often along L than along other
directions. The efficiency of the molecular orientation along L
can be represented by the Legendre polynomials. The order parameter S, which is called the degree of orientation, Φ [Chap.
3, Eq. (34)], is the lowest order nonvanishing Legendre polynomial
that describes the orientational efficiency of nonpolar nematic
liquid crystals. $S = \frac{1}{2}\langle 3 \cos^2 \theta - 1 \rangle_{time\ av}$, where θ is the
angle between the long molecular axis and the preferred direction

<u>L</u>. S = 0 for ordinary isotropic liquids, and S = 1 for a
perfectly aligned nematic at absolute zero. S values can be
measured for nematics using a variety of techniques [25]. S
usually falls between 0.4 and 0.5 for most compounds near the
nematic → isotropic transition temperature, even though the transition temperature itself may vary by several hundred degrees from
compound to compound. At lower temperatures, S increases
monotonically and often can become as large at 0.7 or 0.8 before
the nematic liquid crystal changes phase. The measured values of
S are in good agreement with the values predicted by the Maier-
Saupe theory [26]. The theoretically calculated temperature
dependence of S is shown in Fig. 2. This is essentially a
universal curve, because to a very good approximation it is applicable to all nematics. Applied electric and magnetic fields of
usual laboratory size have a negligible effect on the value of S.

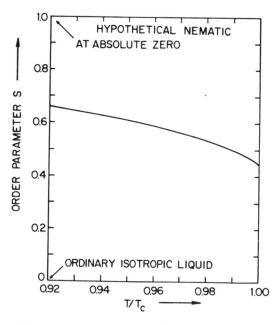

FIG. 2.   Temperature variation of the nematic order parameter
S   for PAA according to the Maier-Saupe theory. This figure can
be regarded as a universal curve for all nematics.

   Elongated molecules that are dissolved in a liquid crystal
solvent will also be oriented to some extent. The solute or
"guest" molecules by themselves need not form a liquid crystalline
phase. The order parameter of the solute molecules will in general be different from the order parameter of the nematic solvent
"host" molecules. Nematic liquid crystals can be used as an
orienting matrix to study the anisotropic spectroscopic properties
of the solute molecules. Saupe [27] has recently reviewed this
field.

2. Anisotropy of Physical Properties

   Since the nematic state is locally uniaxial, many material
properties such as refractive index, dielectric constant, electrical
conductivity, etc. show anisotropic behavior. The magnitudes of
these various anisotropies show a strong temperature dependence which
mainly results from the temperature dependence of the order
parameter S.

   Optically, the known nematic liquid crystals behave like
uniaxial crystals with a positive birefringence; the extraordinary
refractive index n_e is larger than the ordinary refractive index
n_0. The local optic axis coincides with the local preferred
direction L. The birefringence of nematic liquid crystals can be
appreciable. MBBA, for example, is typical, having a birefringence of +0.19 at 25°C for light of 5892 Å wavelength [28,29].

   The low-frequency dielectric anisotropy $\Delta\varepsilon = \varepsilon_1 - \varepsilon_2$ on the
other hand, can be either positive or negative. Here ε_1 and ε_2
are the dielectric constants measured parallel and perpendicular
to the local optic axis, respectively. For a perfectly ordered
nematic (S = 1) these would correspond to the values measured
parallel and perpendicular to the long molecular axis. The sign
of $\Delta\varepsilon$ plays an important role in the mechanism for orientation
and electro-optical behavior in the liquid crystal. The sign and
magnitude of $\Delta\varepsilon$ depend largely upon the direction and magnitude
of the permanent dipole moment in the nematic liquid crystal

molecule [30]. In general, molecules where the dipole moment is
rather large and is directed mainly perpendicular to the long axis
will form a nematic phase with a negative $\Delta\epsilon$, and molecules having a large component of the dipole moment parallel to the long
axis will produce a nematic phase with a positive $\Delta\epsilon$. The
permanent dipole moment of PAA, for example, makes a large angle
with the long molecular axis and $\Delta\epsilon = -0.14$ at 130°C [30].
Nematic liquid crystals composed of molecules having the highly
polar $-C\equiv N$ group in the terminal position have large positive
dielectric anisotropies in the range of 10-20 [31].

ϵ_1 and ϵ_2 also show entirely different frequency depen-
dencies. ϵ_2 behaves like the dielectric constant of an ordinary
polar liquid and shows the normal Debye dispersion in the micro-
wave region. ϵ_1, however, can exhibit an additional dispersion
in the radio frequency region. In fact, this additional dispersion for one compound occurs at 10 kHz where $\Delta\epsilon$ actually changes
sign [32]. This phenomenon is discussed in detail by Martin et
al. [33].

The electrical conductivity of nematic liquid crystals is
also anisotropic. For typical cases the conductivity parallel to
the optic axis is higher than the conductivity perpendicular to
it, although recently Rondelez [34] has discovered some exceptions
to this behavior. As would be expected, the charge carriers must
encounter less hindrance by moving in a direction parallel to the
long axis of the molecules. The ratio of the conductivities
measured parallel and normal to the optic axis σ_1/σ_2 generally
falls in the region 1.3-1.6 for most nematics [35-37], even though
the actual magnitude of the conductivity may vary over several
orders of magnitude. The conductivity of nematic liquid crystals
depends mainly upon the density of charge carriers, which are
usually easily ionizable impurities. The conductivity of MBBA,
for example, can vary from 10^{-6} to 10^{-13} Ω^{-1} cm^{-1} depending upon
its purity.

The space charge limited dielectric relaxation frequency f_0 plays an important role in the understanding of electro-optical effects in liquid crystals. $1/f_0$ is the time constant for the liquid crystal to reach electrostatic equilibrium after the sudden application of an electric charge or potential. In reality, this approach to equilibrium cannot be described by a single time constant because nematic liquid crystals have anisotropic conductivities and dielectric constants. For our purposes, however, it is acceptable to neglect this anisotropy, and we define

$$f_0 = \frac{\bar{\sigma}}{\varepsilon_0 \bar{\varepsilon}}$$

where $\bar{\sigma}$ and $\bar{\varepsilon}$ are the spatial averages of the conductivity and the dielectric constant. The electro-optical effects observed in nematics are quite different depending upon whether the applied electric field has a frequency higher or lower than f_0.

3. Elastic Continuum Theory

In the continuum picture of liquid crystals, first proposed by Zocher [38], the local optic axis $\underline{L}$ can continuously and smoothly change from place to place in a macroscopic sample. It requires a certain amount of energy to make a distortion of the local optic axis, because there are elastic forces present that tend to maintain a uniformly oriented configuration. The elastic energy density g associated with a deformed, nematic sample can be expressed as [39-41]

$$g = \frac{1}{2} k_{11} (\text{div } \underline{L})^2 + \frac{1}{2} k_{22} (\underline{L} \cdot \text{curl } \underline{L})^2 + \frac{1}{2} k_{33} (\underline{L} \times \text{curl } \underline{L})^2$$

$$(1)$$

k_{11}, k_{22}, and k_{33} are the curvature elastic constants corresponding to the respective splay, twist and bend distortions of $\underline{L}$ in space as is illustrated in Fig. 3. These elastic parameters

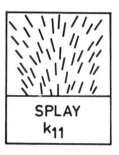

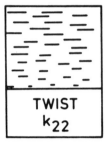

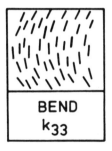

FIG. 3.   The fundamental elastic deformations of a nematic
liquid crystal.

are material constants and have been measured for many nematic

liquid crystals [42-45].   Typical values fall in the neighborhood

of 10^{-5} to 10^{-6} dyn.

Macroscopic samples with a uniform orientation of $\underline{L}$ rarely

form spontaneously in the laboratory because of wall orientation

effects. $\underline{L}$ is generally very rigidly anchored in various direc-

tions at different places on the solid surface interface.   These

binding forces are much stronger than the orienting forces between
the nematic molecules and therefore the bulk of the liquid crystal
is usually distorted in a complicated way. Unidirectionally
rubbing the boundary surfaces with cleaning tissue or fine abra-
sives are proven methods to obtain uniform boundary orientations
[46,47a]. Berreman [47a] has observed submicroscopic parallel
grooves in the boundary surface after rubbing and shows that the
elastic distortion energy of the nematic liquid crystal is minimized when the local optic axis is aligned parallel to the grooves.
Janning [47b] has reported on the excellent liquid crystal orienting properties of thin films that have been vacuum-deposited on a
substrate at oblique angles. Magnetic fields have also been used
to uniformly align nematic liquid crystals. With some nematic
liquid crystals it is possible to uniformly orient the optic axis
perpendicular to the solid boundary surfaces. Saupe [42] obtained
this perpendicular or homeotropic orientation with PAA by treating
the surfaces with hot sodium dichromate-sulfuric acid cleaning
solution. MBBA spontaneously adopts the perpendicular orientation
to surfaces that are smeared with a thin layer of lecithin [48].
Special dopants [49,50] and special surface treatments [51,52]
have also been used to induce homeotropy. By these procedures it
is possible to obtain "single crystal" nematics that are either
parallel or perpendicularly oriented to the boundary surfaces.
Such uniformly oriented samples are required before meaningful and
reproducible electro-optical observations can be obtained.

B.   Cholesteric Liquid Crystals

This type of liquid crystal is called cholesteric because the
first materials discovered to show this mesophase were esters of
cholesterol. Actually, a great number of noncholesterol-derived
molecules are now known to exhibit this mesophase. The only thing
these molecules have in common with cholesterol esters is that

they are all chiralic molecules, i.e. they are nonsuperimposable
with their mirror images.

The cholesteric phase is essentially uniaxial like nematic
liquid crystals except that the local optic axis now uniformly
twists in one direction of space in the undisturbed equilibrium
structure as is illustrated in Fig. 1b. The degree of order can
still be specified by a single order parameter S which is
comparable in magnitude to the value of S for nematics [53].
All that we have discussed in connection with the local uniaxial-
ity of nematics still applies to cholesterics. Alignment at a
boundary surface can still be achieved by rubbing, dielectric
anisotropies can be either positive or negative, etc.

The repeating distance of the cholesteric helicoidal structure is defined to be the pitch P_0. This is the distance measured along the helical axis for $\underline{L}$ to twist around by one
complete turn. P_0 typically falls between 2000 $\overset{\circ}{A}$ and infinity
for most cholesteric compounds. An infinite pitch means that the
structure is nematic. Friedel [18] observed that a cholesteric
liquid crystal with an adjustable pitch could be made by adding
varying amounts of a chiralic substance to a nematic solvent. For
some cholesterics the pitch shows an extreme sensitivity to
temperature that is caused by pretransitional effects from a
nearby smectic mesophase. This sensitivity is useful for some
applications, but for electro-optical devices a nearly constant
pitch is desired over a wide range of temperatures. There are
many cholesteric liquid crystals, however, with temperature
insensitive pitches.

When the pitch of the helical structure is comparable to the
wavelength of visible light, cholesteric liquid crystals show
vivid selective reflection colors. This effect can be thought of
as arising due to the interference of light waves between the
similarly oriented "layers" having a spacing of $P_0/2$. Different
aspects of this optical problem have been solved in detail by a
number of investigators [53-58]. All other things being equal

the wavelength of maximum reflection is proportional to P_0, and the width of the selective reflection band is proportional to the birefringence Δn, where $\Delta n = n_e - n_0$. This ability to selectively reflect light makes cholesteric liquid crystals particularly promising for electro-optical color display device applications.

C. Smectic Liquid Crystals

The smectic mesophase is more highly ordered than nematic mesophase because it is a layered structure of oriented molecules. As with nematics, the long axis of the molecules comprising the smectic mesophase are parallel to the preferred direction L, but, in addition, the molecular centers are also arranged in a series of equidistant planes, resulting in a periodic modulation of the microscopic density extending over large distances. There is no such density wave in cholesteric liquid crystals, however.

Within the smectic planes the molecular centers show a liquid-like randomness as is illustrated in Fig. 1(c). Of the known liquid crystal structures, the smectic structure probably most closely resembles that of biological membranes [59]. This structure is uniaxial when the molecules are normal to the layers (smectic A), and biaxial if the molecules are tilted at some oblique angle to the layers (smectic C). Smectic liquid crystals generally have much higher viscosities than nematics, and they are harder to align by boundary forces or electric and magnetic fields. Only a few electro-optical studies have been made on smectic liquid crystals.

Smectic liquid crystals were investigated in electric fields as early as 1936 [60]. Vistin and Kapustin [61] found similar electro-optical effects to those that were then known to occur in nematics, if about 10 times the field strength was applied. The possibility of electrohydrodynamic instabilities occurring in

smectic A liquid crystals has been considered by Geurst and
Goossens [62]. The experiments of Carr [63] indicate that the
conductivity anisotropy in smectics is of opposite sign to that in
nematics; the conductivity is lower when measured parallel to the
long molecular axis than when it is measured normal to it. This
is reasonable because the charge carriers would be expected to
experience more difficulty in passing through the layers than moving parallel to them. Recently Tani [64a] has reported an
electro-optical storage effect in a smectic liquid crystal which
is characterized by a change in optical transmission upon application of a voltage and which remains after the voltage is removed.
Return to the original state is achieved by heating the sample.
Kahn [64b] has applied a similar effect in a display scheme already developed for cholesteric liquid crystals by Melchior et al.
[65]. Electro-optical studies in smectic liquid crystals are
still in an exploratory stage and many of the results are still
quite speculative. We will therefore not discuss smectic liquid
crystals any further.

II.   ELECTRO-OPTICAL EFFECTS
RESULTING FROM ELECTROHYDRODYNAMIC PROCESSES

In this section we will discuss electro-optical phenomena in
nematic liquid crystals that depend upon electrical conduction and
fluid flow. The deformations that result under these conditions
are sometimes called anomalous deformations, because the optic
axis can align itself in a direction contrary to that expected
from the electric torques. Electro-optical effects resulting from
anomalous deformations are also known to occur in cholesterics
[66-69] and smectics [61] but the theories developed to explain
the effects observed in these mesophases are still at a
comparatively early stage.

A.   Theoretical Background

In the continuum picture, the dynamic behavior of a nematic
liquid crystal is completely specified by the orientation of the
local optic axis $L(r,t)$ and the local fluid flow vector $v(r,t)$.
In principle it is possible to obtain these two vector fields L
and v from the two equations of motion. The first equation requires a balancing of elastic, electric, viscous, and inertial
forces. The second equation of motion requires the balancing of
all internal torques. Leslie [70] and Ericksen [71] were the
first to develop these equations, but their formulations are
difficult to follow. A clearer picture can be obtained from
several recent publications dealing with continuum hydrodynamics
of liquid crystals where the complexities are greatly reduced
[72-75]. In general the equations of motion depend upon quantities such as the electric field strength, space charge density,
dielectric displacement, etc., and therefore a number of additional constitutive equations are needed before the basic equations of
motion can be compared with experiment. Generally, Ohm's law and
a linear relationship between the electric field and the dielectric displacement are assumed for a constant deformation of the
liquid crystal. At least four different kinds of hydrodynamic
instabilities have been observed in nematic liquid crystals.
These instabilities will be discussed in detail in the following
sections in terms of the most recent developments in the continuum
theory.

B.   Williams Domains

One of the most striking electro-optical effects displayed
by liquid crystals is a periodic striped pattern referred to as
Williams domains [76], that appear in many nematic liquid crystals

in the parallel-oriented, sandwich-cell geometry. This pattern becomes visible if transparent electrodes are used in this geometry making it possible to look through the electrodes in a direction parallel to the electric field. Figure 4 shows a photomicrograph of the striped pattern that is observed in transmitted light through the film. The transparent electrodes are made from glass upon which a thin, conducting layer of SnO_2 has been deposited. To make a sandwich cell, a nematic liquid crystal is allowed to flow between two such conducting surfaces that are parallel and spaced about 10-100 µm apart. The electrodes are specially treated to make the optic axis of the nematic liquid crystal uniform and parallel to both surfaces.

Williams domains appear in the sample only above a certain threshold voltage that is independent of sample thickness. For many nematic materials this threshold voltage falls in the neighborhood of 4-8 V. The periodicity of the striped pattern is roughly 1.2-1.4 times the thickness of the sample and is nearly independent of the applied voltage [76,77]. Williams domains are

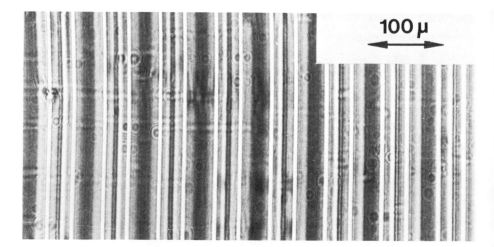

FIG. 4. Virtual focal lines of the Williams domains in an MBBA layer. (Photograph kindly furnished by J. Nehring, Brown Boveri Research Center, Baden, Switzerland.)

associated with a cellular fluid flow which can be clearly seen
from the periodic motion that the flow induces in suspended dust
particles [77,78]. As the voltage is increased above the threshold value the flow and the deformation are observed to become
stronger and at higher voltages the Williams domains become unstable; the parallel stripes develop more and more bifurcations
and the stripes themselves begin to move. Finally at still higher
voltages a general turbulence sets in that strongly scatters
light. This process is referred to as dynamic scattering and will
be discussed further in Sec. IIE.

This periodic deformation is easily seen in unpolarized light,
but the pattern appears to vanish when viewed with polarized light
if the electric field vector of the incident light is perpendicular to the direction of the original nematic alignment. This
behavior indicates that the local optic axis in the deformed
nematic layer always lies in the plane defined by the applied
electric field and the original nematic alignment. This plane is
indicated in Fig. 5.

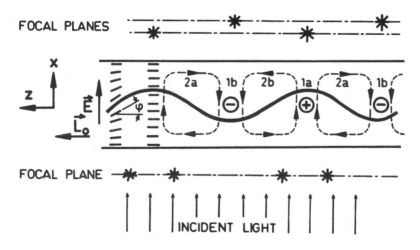

FIG. 5. Schematic illustration of Williams domains showing
the interrelation of the cellular fluid flow (dashed lines), the
deformed optic axis (heavy line), the focal lines (asterisks),
and the space charges (+ and -).

Since nematic liquid crystals are birefringent, a great deal can be learned about the deformation by observing the sample in monochromatic light between crossed polarizers. The optical path-length difference between the ordinary and extraordinary rays passing through the sample strongly depends upon the deformation. The periodic pattern of the Williams domains therefore appears as regularly spaced groups of interference fringes that indicate where the maximum deformation occurs. By combining these inter-ference measurements with observations on the velocity of trapped dust particles, it is found that the liquid crystal flows perpendicularly to the electrodes in the regions where the defor-mation is at a minimum and parallel to the electrodes where the deformation is at a maximum. These observations are sketched in Fig. 5 where the dashed lines with arrows indicate the direction of the cellular fluid flow. It should be emphasized that the deformation itself is static even though it is produced by a material flow. The orientation of the local optic axis in the laboratory coordinate system is fixed, whereas it shows a periodic time dependence in a coordinate system moving with the fluid.

Under the microscope in transmitted parallel light the Williams domains appear as sharp, bright lines at three different voltage-dependent focal planes, which all lie outside the volume of the sample. The periodically deformed nematic behaves like an arrangement of alternating converging and diverging cylindrical lenses [77,79]. Since nematic liquid crystals are locally uniaxial, two completely independent light rays can propagate in-side the material. A deformation in the sample does not affect propagation of the ordinary ray, but it will have a large effect upon the extraordinary ray. A periodically deformed sample will therefore present different effective refractive indexes to extraordinary rays in different parts of the sample. The deforma-tion monotonically increases as we move from region 1 to region 2 in Fig. 5, for example, which results in a progressively smaller

effective refractive index being presented to a normally incident
light beam. The region between 2a and 2b will therefore act like
a converging lens and the region between 1a and 1b will behave
like a diverging lens. The detailed solution to this optical
problem is difficult, because the direction of the wave normal
generally does not coincide with the direction of wave propagation.
Nehring [80] has shown that there are really two types of converging lenses and two types of diverging lenses. The real focal
lines of converging lenses are equally spaced and lie above the
sample in the two different planes indicated by the asterisks in
Fig. 5. The virtual focal lines of the diverging lenses have
unequal spacing and lie in the same plane below the sample.
Nehring has verified these predictions by experiment. The
unequally spaced lower, virtual focal lines are seen on the
photomicrograph in Fig. 4.

This arrangement of periodic lenses acts like a diffraction
grating. A number of investigators have studied the diffraction
patterns of a light beam normally incident to a nematic sample
showing Williams domains [81]. Carroll [82] has recently calculated the intensities of the various diffraction orders that would
be expected from such a grating and finds good agreement with
experimental intensities.

According to the Carr-Helfrich [83,84a] model, Williams
domains arise because the nematic liquid crystal has an aniso-
tropic electrical conductivity. For most cases $\Delta\sigma = \sigma_1 - \sigma_2 > 0$
[35-37], where σ_1 and σ_2 are the electrical conductivities
measured parallel and perpendicular to the local optic axis,
respectively. It follows, then, that if a parallel oriented
nematic layer becomes infinitesimally deformed due to thermal
fluctuations, for example, a component of electric current will
flow in the z direction parallel to the electrodes, creating
real space charge in the sample. This space charge can in turn
produce an infinitesimal fluid flow which will, through viscous
forces, contribute to the torques acting upon the local optic

axis. At a certain threshold electric field $E_{0\sigma}$, which depends upon the periodicity λ of the deformation, the flow-induced torque will be large enough to overcome the infinitesimal electric and elastic torques which tend to maintain the uniformly oriented configuration. Above this field the infinitesimal deformation will therefore be reinforced and a real fluid flow will occur in the material. This process can be understood by examining the dynamic behavior of infinitesimal deformation in the sample.

The continuum equations referred to in Sec. IIA are greatly simplified for small deformations because it is possible to make the linear, small-angle approximation. Furthermore, if it is assumed that the fluctuation and flow have spatial periodicities λ in the z direction (Fig. 5) and an exponential time response to a stepwise change in voltage, then the response time τ for such a fluctuation is [84b]

$$\frac{1}{\tau} = \frac{4\pi^2 k_{33}}{\gamma^2 [2\alpha_2^2/\{\alpha_4 - \alpha_2 + \alpha_5\} + (\alpha_2 - \alpha_3)]} \left[\left(\frac{E}{E_{0\sigma}}\right)^2 - 1 \right] \qquad (2)$$

where $1/\tau < f_0$, k_{33} is the bend elastic constant (Sec. IAc), and the α_i (I = 1,5) are the five independent viscosity parameters defined by Leslie [85]. $E_{0\sigma}$ is the dc threshold field for the instability which has been calculated by Helfrich [84a].

$$E_{0\sigma} = \frac{2\pi}{\gamma} \left\{ \frac{k_{33}}{\varepsilon_0 \varepsilon_1 \{(\Delta \varepsilon \nabla_2 / \varepsilon_1 \nabla_1) - [2\alpha_2/(\alpha_4 - \alpha_2 + \alpha_5)][\varepsilon_2/\varepsilon_1) - (\sigma_2/\alpha_1\}} \right\}^{\frac{1}{2}} \qquad (3)$$

When $E < E_{0\sigma}$, τ is negative and the fluctuation exponentially damps out with time. When $E > E_{0\sigma}$ τ is positive and the intensity of the fluctuation grows exponentially with time. Of course, the growth is exponential only as long as the deformation angle is small in comparison with its final saturation value. $E_{0\sigma}$

therefore represents a true threshold field.

The periodicity λ of the Williams domains is observed to be proportional to the sample thickness, x_0. This observation allows us to transform Helfrich's expression (3) for a threshold field into an expression for a true threshold voltage $U_{0\sigma}$ which is independent of sample thickness.

$$U_{0\sigma} = 2\pi \frac{x_0}{\tau} \left\{ \frac{k_{33}}{\varepsilon_0 \varepsilon_1 \{(\sigma_2 \Delta \varepsilon_1 \sigma_1) - (2\alpha_2/[\alpha_4 - \alpha_2 + \alpha_5][\varepsilon_2/\varepsilon_1 - \sigma_2/\sigma_1])\}} \right\}^{\frac{1}{2}} \quad (4)$$

For PAA, $\lambda \cong 1.3\ x_0$ [86] and the cellular flow pattern describes an ellipse. Inserting this and the latest material constants [77] measured for PAA into Helfrich's threshold voltage expression (4) gives $U_{0\sigma} = 5.6 \pm 1$ V. This is in reasonable agreement with the measured value of 8.3 ± 0.3 V [87] considering the large uncertainties in many of the material constants and the nature of the assumptions.

Helfrich's theory cannot predict the actual wavelength for the Williams domains because it is based on a one-dimensional model where the deformation angle ϕ varies only along the z direction as is illustrated in Fig. 5. His model therefore neglects the presence of the boundaries where the deformation angle must be zero.

Penz and Ford [72] have extended Helfrich's theory into two dimensions with a series of numerical calculations on PAA and MBBA. Their results unambiguously predict the existence of a threshold voltage and that the wavelength of the instability is proportional to the layer thickness. Such predictions can only be made from a two-dimensional theory that explicitly considers the boundaries. Their numerical results indicate a threshold voltage of 8.2 V with $\lambda = 1.42\ x_0$ for PAA, in fair agreement with experiment. It is difficult to estimate the effect of uncertainty in the material parameters on the numerical calculations. A more precise comparison between theory and experiment will have to wait for more accurate material values.

For voltages above the threshold value, Penz and Ford's [72] numerical results map out regions of instability for the nematic liquid crystal. Their calculations show that for a given applied voltage an instability can occur only if the wavelength of the disturbance falls within a prescribed region. Because their calculations are based upon the linear, small angle approximation, however, it is not possible to predict which wavelength in this region will actually occur; only at the threshold voltage is the wavelength for the instability single-valued. Nevertheless, it is interesting to speculate whether Williams domains fall near the short or long wavelength limit of the instability. The computed long wave limit is practically independent of applied voltage and it is tempting to associate Williams domains with this limit. At the short wavelength limit, however, the computed wavelength of the instability is inversely proportional to the applied voltage and this bears a striking correspondence to the observations of Vistin' [88], and Greubel and Wolff [89]. Using unusually thin nematic specimens $(x_0 \leq 6 \ \mu m)$ having a moderately low conductivity $(\sigma = 2 \times 10^{-11} \ \Omega^{-1} \ cm^{-1})$ Greubel and Wolff report that the periodicity of the striped pattern is inversely proportional to the applied electric field up to the highest practicable dc field strengths. Nonlinear terms must be included in Penz and Ford's treatment before it will be possible to predict the actual wavelength for an instability above the threshold voltage.

Helfrich's [84a] treatment of Williams domains is strictly valid only for dc fields in the steady state where the space charge density is independent of time. Dubois-Violette et al. [75] have extended Helfrich's treatment to include the case for alternating field excitation. They also begin with the small angle approximations to the continuum equations referred to in Sec. IIA but allow for the time dependence of the space charge. For the case when the frequency of the applied voltage is much

lower than the space charge limited dielectric relaxation
frequency, the formation of space charge is able to follow the
voltage and the expression for the threshold field is given by

$$\langle E(t)^2 \rangle_{\text{time av}} = E_{0\sigma}^2 \frac{(\zeta^2 - 1)(1 + \omega^2 \tau_0^2)}{\zeta^2 - (1 + \omega^2 \tau_0^2)} = E_{0\sigma}^2 \frac{\omega_c^2 (1 + \tau_0^2 \omega^2)}{(\omega_c^2 - \omega^2)}$$

(5)

where $1/\tau_0 \approx f_0$ (Sec. IA2) and $\omega\tau_0 \ll 1$. ζ^2 is the
"Helfrich" parameter characteristic of the liquid crystalline
material and ω is the frequency of the excitation voltage. The
deformation itself is still stationary even though an alternating
field is applied.

For dc fields this expression reduces to Helfrich's threshold
equation (3). Equation (5) predicts a cutoff frequency ω_c
where the threshold field becomes infinitely large. By assuming
that the period of the deformation is proportional to the specimen thickness, as was done for the dc case, Eq. (5) predicts a
threshold voltage that is independent of sample thickness. The
inset of Fig. 6 shows the threshold field (actually a threshold
voltage) that was measured for the formation of Williams domains
in MBBA. The curve drawn through the data points is calculated
from Eq. (5) with $\zeta^2 = 5.27$ and $\tau_0 = 5.68 \times 10^{-3}$ sec. The
experimental cutoff frequency is 58 Hz.

C.   Chevron Pattern

The Dubois-Violette theory predicts that an entirely
different distortional process will occur in nematic materials
with $\Delta\varepsilon < 0$ at frequencies above f_0. The new pattern that is
observed has a herringbone appearance based on a chevron motif.
It is therefore referred to as the "chevron" pattern [90,91]. A
photograph of this pattern is shown in Fig. 7. Careful observation shows that the amplitude of the chevron deformation follows

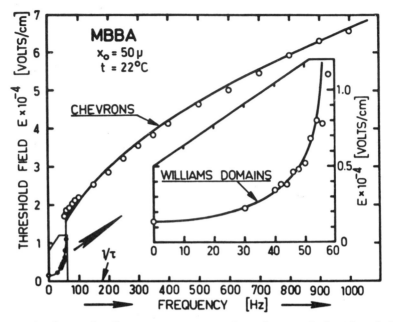

FIG. 6. Main figure: Frequency dependence of the threshold
field for chevrons. Inset: Frequency dependence of the threshold
value for Williams domains.

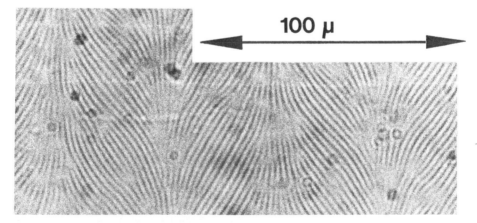

FIG. 7. Chevron pattern observed in a 50 μm thick MBBA layer
for an applied voltage of 70 V at 150 Hz. The cutoff frequency is
58 Hz.

the ac field at double the exciting frequency, in contrast to the
Williams domain pattern which is stationary in ac fields.
Measurements on samples of various thicknesses indicates that the
deformation is characterized by a threshold field rather than a
threshold voltage.

These experimental observations can be understood within the
framework of the Dubois-Violette et al. theory. At frequencies
above f_0 the space charges in the sample cannot follow the frequency. It is possible, however, for the deformation itself to
follow the frequency if a large enough field is applied, because
the response time for molecular alignment is proportional to E^{-2}
(see Sec. IIIB). This type of response to a strong high frequency
electric field is called the "fast turn off mode" [92]. A portion
of the chevron deformation is schematically illustrated in Fig. 8
for a half-cycle of the exciting field. In the steady state,
space charge is trapped at the positions indicated. For the first
part of the cycle shown in Fig. 8a where $E = 0$ the fluid flow is
nearly zero (as indicated by the small arrows), and the deformation is maximum. When the electric field is at its maximum, as is
illustrated in Fig. 8b the fluid flow is also at its maximum, and
the deformation is zero. The entire process then reverses itself
over the next half cycle shown in Fig. 8c and so on.

The Dubois-Violette et al. [75] theory predicts that the
chevron instability should occur at a definite threshold field
which is proportional to $\omega^{1/2}$. These predictions are verified by

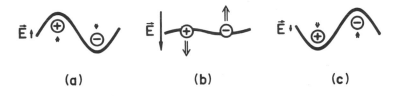

(a)   (b)   (c)

FIG. 8. The chevron deformation. Its amplitude is plotted
at various phases of the excitation voltage and the double arrows
indicate the fluid flow intensity.

experiment. Figure 6 shows the experimental frequency dependence
of the threshold field for the chevron pattern compared with the
theoretical curve.

In contrast to Williams domains, the periodicity of the
chevron pattern depends mainly upon the material constants of the
particular liquid crystal. The period generally falls in the
neighborhood of several micrometers and is insensitive to sample
thickness. According to the basic Dubois-Violette et al. theory,
the spatial period of the chevron pattern should be proportional
to $\omega^{-1/2}$. This model, however, neglects the effect of ionic
diffusion currents between the space charges that become important
at higher frequencies where the spatial period is small. The re-
fined model of Galerne et al. [93] specifically considers the
effect of ionic diffusion on the spatial period and is in good
agreement with experimental measurements except for a high frequency anomaly. Dubois-Violette et al. [75] show that the chevron
periodicity should depend upon the field strength of a magnetic
field applied along the electric field direction or upon an
additional high frequency electric field applied along the same
direction. This effect has been experimentally verified for both
magnetic [91] and electric [94] fields.

In a theoretical paper, Dubois-Violette [95] has made a
series of numerical calculations of the threshold voltage in the
region immediately above the cutoff frequency where analytical
procedures become too complex. She predicts that the threshold
voltage vs. frequency curve should have a pronounced S shape
when the sample is excited with a succession of rectangular
pulses. This behavior has been experimentally observed by the
Orsay Liquid Crystal Group [96]. The large difference observed in
the threshold fields measured for sine and square wave excitation
of the same frequency [96,97] can be explained by the high frequency harmonic content of the square waves.

D. Direct Current Instability

We have seen that Williams domains are characterized by a
cellular fluid flow and can only occur in a medium having an
anisotropic electrical conductivity. These domains are observed
with dc excitation fields as well as with ac fields having frequencies extending into the kilohertz range, depending upon the
sample conductivity. In the isotropic liquid phase the Williams
domains vanish and there is no motion of suspended dust particles,
indicating the absence of cellular fluid flow.

There is another electro-optical effect occurring in nematics
at dc or very low frequency ac fields ($\leq$ a few hertz) that often
appears similar to Williams domains [98]. The threshold voltage
for this instability is comparable to that measured for Williams
domains, but this instability, as was first observed by Koelmans
and Van Boxtel [99], is characterized by a cellular fluid flow
that persists even in the isotropic phase. The threshold voltage
in the isotropic phase is very nearly the same as in the nematic
phase and the cellular flow can be identified by flow birefringence and dust particle motion. Clearly, this effect has nothing
to do with the anisotropic character of the nematic phase.

In ordinary liquids such as benzene, for example, a dc
instability involving cellular flow is known to occur. Felici
[100] explains this effect as arising from unipolar charge injection from the electrodes. A fluctuation of injected charge density or an irregularity of the electrode surface leading to a
nonuniform charge injection will induce fluid flow and build up a
cellular motion. The measured and computed threshold values for
nitrobenzene are in good agreement with Felici's theory. For PAA
and MBBA, however, the measured threshold voltage of 3-5 volts is
much less than the value of about 30 V predicted from theory
[101]. As de Gennes [102] points out, there is clearly a need for

more experiments to determine the nature of the electrochemical
processes occurring at the electrodes and to assess the relative
importance of injected and intrinsic charge carrier mechanisms.

The spontaneous birefringence of the nematic phase makes any
fluid flow readily visible. The dc instability flow pattern
typically appears as an arrangement of irregular polygonal cells
[103]. If the threshold voltages for the Williams domains and the
dc instability are nearly the same the two fluid flows will inter-
fere and many different kinds of optical patterns are possible.
For this reason, it is better to study Williams domains with an ac
applied field.

E. Dynamic Scattering

The effects of an electric field on the light transmission of
a nematic liquid crystal was first described in the pioneering
work of Björnståhl in 1918 [104]. He describes Williams domains,
dielectric alignment of nematics (Sec. IIIA,B) and dynamic scattering. His experiments were made at a time when room temperature
nematic materials and suitable transparent electrodes were not
available and the technical applicability of these effects seemed
remote. 50 years later a paper by Heilmeier, Zanoni and Barton
[105] gave an impetus to liquid crystal research and attracted
world-wide interest because it demonstrated the feasibility of
liquid crystal electro-optic display devices. The dynamic scattering mode described in this paper is still the most often used
liquid crystal electro-optical effect for technical applications.

The dynamic scattering mode is achieved when the voltage is
increased a certain amount beyond the threshold for the formation
of Williams domains where a very turbulent state is reached which
strongly scatters light. There appears to be no true threshold
voltage for this effect but usually about 30 volts is sufficient
to induce strongly turbulent flow [106].

The local optic axis of the nematic liquid crystal makes very complicated contortions in this state of turbulent fluid flow. Any region where the optic axis changes direction acts like a lens because of the birefringence of the nematic, as was already discussed in connection with Williams domains. In the dynamic scattering mode these lenses are smaller and stronger and lead to a coherent forward scattering of an incident light beam. With comparatively high applied voltages, the incident light is depolarized and the scattering cone has rotational symmetry even if the sample was originally in a uniformly ordered state. At lower voltages the scattering angle is less and begins to show an azimuthal angular dependence with respect to the original nematic alignment. Laser diffraction studies of a nematic liquid crystal in the dynamic scattering mode reveal that there is still a remarkable amount of hidden structure present in the turbulent flow [81]. The diffraction patterns obtained with ac excitation are quite different in appearance to those obtained with dc excitation.

At present, there has been comparatively little theoretical work done on dynamic scattering. Penz and Ford [72] speculate that a higher threshold voltage computed for a hydrodynamic solution consisting of two parallel layers of vortices may represent the threshold for dynamic scattering. They surmise that one solution will try to dominate the other and in the process create the turbulence known as dynamic scattering. Unfortunately, these ideas cannot be made more concrete in the framework of a linear theory. A solution to the actual hydrodynamics of the dynamic scattering mode is a difficult problem because the method of making infinitesimal deformations which was very helpful in calculating the threshold values for Williams domains and chevrons cannot be applied here where the deformations are large. By including third-order terms of the infinitesimal deformation in the torque balance equation of the Leslie-Ericksen continuum theory (Sec. IIA), Carroll [107] has recently been able to extend the Helfrich

theory to cover the region above the Williams domain threshold
where he derives the voltage dependence of the amplitude of the
deformation.

By considering higher order terms of the infinitesimal
deformation in the force and torque balance equations, Gruler
[84b] has shown that the striped pattern becomes increasingly unstable at higher voltages. This instability occurs because a
phase shift is introduced that prevents the positions of maximum
fluid flow from coinciding with the positions of minimum deformation. The end result is that the whole pattern of Williams
domains starts to shift in one direction. This can be interpreted
as the beginning of the dynamic scattering mode. This phase shift
is small for thin samples and dynamic scattering should be suppressed. This is consistent with the observations of Vistin' [88]
and Greubel and Wolff [89]. In thick samples, even though there
will be a large phase shift, the motion of the Williams domain
pattern will be sluggish. It should therefore be possible to
optimise the dynamic scattering mode with respect to sample thickness.

The various technical applications of dynamic scattering has
markedly grown since 1968. Only a few of the recent applications
will be briefly mentioned here. A number of investigators have
proposed feasible x-y addressed displays [108,109]. The use of
photoconducting electrodes has opened up new possibilities for
light valves, light amplifiers, and image converters [110]. The
dynamic scattering mode also has applications in holography [111a].

III.   ELECTRO-OPTICAL EFFECTS
RESULTING FROM PURE DIELECTRIC ALIGNMENT

In this section we will discuss electro-optic phenomena in
liquid crystals that do not involve fluid flow in the steady
state. We shall consider only the deformations of a liquid crystal that are produced by a balancing of the elastic and electric

torques acting upon the local optic axis; flow-induced viscous
torques will not be considered. These distortions are called
dielectric alignment or simply "field effect" deformations.
Dielectric alignment distortions in nonconducting materials have
corresponding magnetic analogs where a magnetic field interacts
with the anisotropic diamagnetic susceptibility of the liquid
crystal. For the magnetic analogs there can be no conduction
effects to complicate the situation and it is possible to obtain
the deformation without any further assumptions. The general
problem of dielectric alignment in a conducting nematic liquid
crystal, however, can be solved only if the mechanism for producing the charge carriers is known [111b]. We will restrict our
treatment here to the case where the effects of space charge are
negligibly small. This condition can be realized in very pure
materials or when the frequency of the applied field is much
higher than the space charge limited dielectric relaxation frequency f_0 (Sec. IA2).

Depending upon the kind of liquid crystal chosen and the
optic axis orientation with respect to the electrodes, many types
of distortions with their associated electro-optical behavior can
be observed. In the following section we will deal with some of
these effects that are of theoretical interest as well as of
interest for constructing practical electro-optical devices.

A. Parallel Oriented Nematic Layer

Consider the sandwich cell geometry illustrated in Fig. 9a
where a nematic liquid crystal is confined between two parallel
electrodes having a spacing x_0. By special treatment of the
electrodes (Sec. 1A3) the optic axis of the nematic can be
rigidly anchored along the z direction at the boundaries. This
is the same specimen geometry that has been used to study the
Williams domains described in Sec. IIA. Optical observations can

be made in transmitted light through the nematic material in a
direction perpendicular to the transparent, but electrically conducting electrodes. For nematics with $\Delta\varepsilon > 0$ this parallel
oriented configuration can become unstable when a potential is
applied to the electrodes, because there will be an electric
torque tending to make the optic axis perpendicular to the plates.
Of course, there must also be an elastic torque tending to restore
the system to the undeformed planar configuration. Since all
forces are conservative in this problem, we can determine the
deformation in an electric field from a variational treatment of
the elastic distortion energies and electrostatic energies of the
system.

The magnetic analog of this problem was observed by several
early workers [112] and a theoretical treatment was given, first
by Saupe [42] and later, a more rigorous treatment was given by
Dafermos [113]. Both these treatments assume that the local
optic axis of the nematic is rigidly anchored at the boundaries.
Rapini and Papoular [114] have also considered the case where the
boundary forces are weaker, giving the local optic axis the additional freedom to reorient itself at the boundary surfaces. The
solution for the deformation in a nonconducting nematic in an
electric field, however, is not strictly analogous to the magnetic
case because $\Delta\varepsilon$ is not negligible compared to ε_1 and ε_2 for
most nematic liquid crystals. Therefore, the assumption corres-
ponding to a uniform magnetic field in the deformed nematic layer
is not valid for electric fields. The electric field inside the
deformed sample can show large nonuniformities [115]. A critical
comparison of the deformation produced by electric and magnetic
fields is given elsewhere [43]. In an electrically conducting
liquid crystal the dielectric alignment deformation can be solved
in the weak distortion limit under the assumption that there is no
space charge present in the undeformed layer [111b]. The threshold voltage remains constant in this simple treatment, but above
threshold the strength of the deformation depends upon the

frequency of the applied field. For frequencies above $3f_0$,
however, the nonconducting limit is reached and the effects of
space charge in the deformed layer can be neglected. The solution
that we shall summarize here is in this limit.

Above a certain threshold voltage [116a] $U_{0\varepsilon}$ the theory
predicts that the parallel-oriented nematic layer will distort as
is shown on the right-hand side of Fig. 9a. Below this critical
voltage there is absolutely no distortion of the layer. This
threshold voltage is independent of sample thickness and is given
by

$$U_{0\varepsilon} = \pi \left(\frac{k_{11}}{\varepsilon_0 \Delta\varepsilon} \right)^{1/2} \tag{6}$$

where k_{11} is the splay elastic constant (Sec. 1Ac) and $\Delta\varepsilon$ is
the dielectric anisotropy (Sec. IA2). There is a two fold
degeneracy in this problem; the optic axis in the center of the
film can either slope to the right or tilt by an equal but
opposite amount to the left. Under certain circumstances this
degeneracy can result in the formation of inversion walls in the
layer [116b]. In practice, however, this degeneracy is lifted by
inevitable boundary irregularities and the deformation in the
sample generally tilts in the same direction over large regions.

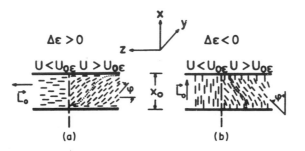

FIG. 9.   (a) Dielectric alignment in a parallel oriented
sandwich cell.   (b) Dielectric alignment in a perpendicularly
oriented sandwich cell.

The maximum deformation angle of the local optic axis ϕ_M occurs in the center of the nematic layer as is illustrated in Fig. 9a. ϕ_M is implicitly related to the applied potential U by [43,116c]

$$\frac{U}{U_{0\varepsilon}} = \frac{2}{\pi}\left(1 + \frac{\Delta\varepsilon}{\varepsilon_2}\sin^2\phi_M\right)^{1/2}\int_0^{\phi_M}\left[\frac{1 + \kappa\sin^2\phi}{(1 + \frac{\Delta\varepsilon}{\varepsilon_2}\sin^2\phi)(\sin^2\phi_M - \sin^2\phi)}\right]^{1/2}d\phi \tag{7}$$

where κ is an elastic constant anisotropy parameter given by $\kappa = (k_{33} - k_{11})/k_{11}$. In the small deformation region immediately above the threshold voltage (7) simplifies to

$$\frac{U}{U_{0\varepsilon}} = 1 + \frac{1}{4}(\kappa + \frac{\Delta\varepsilon}{\varepsilon_2} + 1)\phi_M^2 + 0(\phi_M^4) \tag{8}$$

Figure 10 shows the calculated dependence of ϕ_M on $U/U_{0\varepsilon}$ for $\kappa = 0.7$ and several values of $\Delta\varepsilon/\varepsilon_2$. ϕ_M depends strongly on κ [43] but it is comparatively insensitive to changes in $\Delta\varepsilon/\varepsilon_2$.

Once ϕ_M is known, the deformation angle in the other parts of the sample $\phi(x)$ can be computed from the expression

$$\frac{x}{x_0} = (\frac{1}{2})\frac{\displaystyle\int_0^\phi\left\{\frac{(1 + \kappa\sin^2\phi)(1 + \Delta\varepsilon/\varepsilon_2\sin^2\phi)}{\sin^2\phi_M - \sin^2\phi}\right\}^{1/2}d\phi}{\displaystyle\int_0^{\phi_M}\left\{\frac{(1 + \kappa\sin^2\phi)(1 + \Delta\varepsilon/\varepsilon_2\sin^2\phi)}{\sin^2\phi_M - \sin^2\phi}\right\}^{1/2}d\phi} \tag{9}$$

where x_0 is the sample thickness. Near the threshold potential this expression also simplifies to

$$\frac{x}{x_0} = \frac{1}{\pi}\arcsin\frac{\phi}{\phi_M} - \phi(\phi_M^2 - \phi^2)^{1/2}\frac{1 + 3(\kappa + \Delta\varepsilon/\varepsilon_2) + \cdots}{12\pi(1 + \frac{1}{4}(\kappa + \Delta\varepsilon/\varepsilon_2 + 1)\phi_M^2 + \cdots)} \tag{10}$$

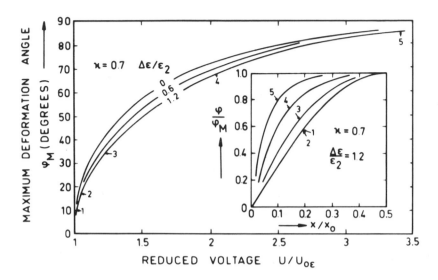

FIG. 10. Main figure: Voltage dependence of maximum deformation angle occuring in the center of the layer. The curve shown for $\Delta\varepsilon/\varepsilon_2 = 0$ is only of theoretical interest because the threshold voltage would be infinite; it is included as a comparison standard since it corresponds exactly to the deformation expected for a magnetic field. Inset: Spatial dependence of the electric field induced deformation for several applied voltages.

Equation (9) is integrated numerically and the results are shown in the inset of Fig. 10 in reduced form for several values of ϕ_M with $\kappa = 0.7$ and $\Delta\varepsilon/\varepsilon_2 = 1.2$. For weak deformations the curve is sinusoidal as is predicted by Eq. (10). For strong deformations where $U \gg U_{0\varepsilon}$ and $\phi_M \approx \pi/2$ the curve becomes nearly square. This indicates that there is still a thin undeformed layer adjacent to each electrode where $\phi \approx 0$. In this limit Fig. 11 is a good approximation to the true deformation. The

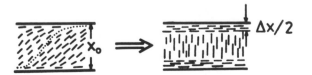

FIG. 11. Approximation for the deformation at high fields.

average deformation angle $\bar{\phi}$ and the undeformed layer thickness $x/2$ are related to the applied potential by

$$U \gg U_{0\varepsilon}$$

$$\frac{\Delta x}{x_0} = \frac{2}{\pi} \frac{U_{0\varepsilon}}{U} = 1 - \frac{2}{\pi} \bar{\phi}$$

$$k_{11} = k_{33}$$

We are now in a position to calculate any change in the properties of the sample brought about by the dielectric alignment. The birefringence of the sample, for example, is strongly influenced by the deformation. Saupe [42] has already computed the birefringence changes in the magnetic analog, and we need only modify his procedure to make it applicable for electric fields. For normally incident light of wavelength λ, the phase difference change, $d(0) - d(U)$, between the ordinary and extraordinary rays passing through the sample before and after the deformation is given by

$$d(0) - d(U) = \left| \frac{x_0}{\lambda} n_e \left\{ 1 - \frac{\displaystyle\int_0^{\phi_M} \left\{ \frac{(1 + \kappa \sin^2 \phi)(1 + \Delta\varepsilon/\varepsilon_2 \sin^2 \phi)}{(1 + \nu \sin^2 \phi)(\sin^2 \phi_M - \sin^2 \phi)} \right\}^{1/2} d\phi}{\displaystyle\int_0^{\phi_M} \left\{ \frac{(1 + \kappa \sin^2 \phi)(1 + \Delta\varepsilon/\varepsilon_2 \sin^2 \phi)}{\sin^2 \phi_M - \sin^2 \phi} \right\}^{1/2} d\phi} \right\} \right|$$

(11)

Making the small angle approximation that is valid near the threshold potential $U_{0\varepsilon}$ we obtain

$$d(0) - d(U) = \left| \frac{x_0}{\lambda} n_e \nu \frac{U - U_{0\varepsilon}}{U_{0\varepsilon}} \frac{1}{\kappa + 1 + \Delta\varepsilon/\varepsilon_2} + 0\left(\frac{U - U_{0\varepsilon}}{U_{0\varepsilon}}\right)^2 \right|$$

(12)

where $\nu = (n_e^2 - n_0^2)/n_0^2$. The phase difference can be "tuned" by

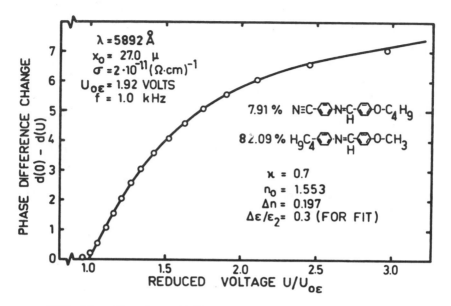

FIG. 12.   The phase difference change between the ordinary
and extraordinary light rays in the deformed, parallel oriented
sample.   The curve is calculated from Eqs. (8) and (11) with
$\kappa = 0.7$ and $\Delta\varepsilon/\varepsilon_2 = 0.3$.

varying the applied voltage   U.   Figure 12 shows the good
agreement between a theoretical curve generated from Eqs. (8) and
(11) and the experimental data points.

The optical absorption of the sample can also be influenced
by the dielectric alignment.   The natural pleochroism exhibited
by many nematic liquid crystals can be greatly augmented by adding
strongly pleochroic dyes as solute molecules which are oriented
by the liquid crystal solvent (Sec. IA1).   The reorientation of
the dye molecules by an electric field induced dielectric alignment will then change the optical absorption of the sample [117].
An experimental optical transmission curve as a function of
applied voltage is given in Fig. 13.

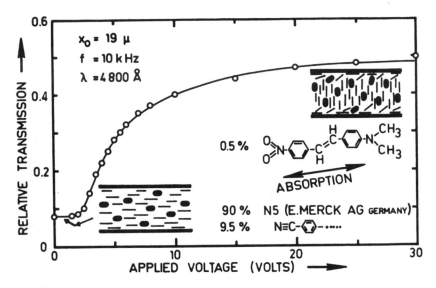

FIG. 13.   Experimental optical transmission curve for a
nematic liquid crystal containing a dissolved pleochroic dye.
(A. Stieb and G. Baur, unpublished data.)

B.   Perpendicularly Oriented Nematic Layer

A second important nematic geometry for electro-optic
applications is the perpendicularly ordered film or homeotropic
geometry.   It was discussed in Sec. 1A3 how it is possible to
anchor the local optic axis of a nematic liquid crystal perpendicular to the boundary surfaces.   This situation is sketched in the
left hand side of Fig. 9b where the upper and lower electrodes are
transparent, but electrically conducting.   If $\Delta\varepsilon < 0$,   a dielectric alignment can occur which is similar to the dielectric alignment in a parallel oriented layer for which $\Delta\varepsilon > 0$.   Above a
critical threshold voltage the nematic layer is deformed as is
sketched on the right hand side of Fig. 9b.   This deformation is
sometimes referred to as the homeotropic Fréedericks transition.
To calculate the actual shape of the deformation it is only
necessary to interchange k_{11} with k_{33}, ε_1 with ε_2, and n_e
with n_0 in the expressions (6)-(12).   The threshold voltage, for

example, is given by

$$U_{0\varepsilon} = -\pi \; \frac{k_{33}}{\varepsilon_0 \Delta\varepsilon} \tag{13}$$

An important difference between this geometry and the parallel
oriented layer of Sec. IIIA is that here there is an infinite fold
degeneracy in the tilt direction of the optic axis in the deformed
nematic layer. In practice, however, this can usually be reduced
to a two fold degeneracy by unidirectionally rubbing the boundary
surfaces.

The homeotropic geometry has recently received much attention
in connection with electro-optic information display devices [51,
118-120]. In one scheme, white light is incident to a
perpendicularly oriented nematic layer having transparent electrodes that is placed between crossed polarizing films. When the
layer is deformed through the dielectric alignment process, the
ordinary and extraordinary light rays leaving the sample will have
a wavelength-dependent phase difference and can selectively inter-
fere at the analyzer to produce a transmitted color. The effective birefringence of the nematic can be voltage-tuned over a wide
range to produce practically any desired color. Unfortunately,
however, the colors are strongly temperature dependent because of
the inherent temperature dependence of the nematic degree of order
S (Sec. IA1). Soref and Rafuse [119] show how the dielectric
alignment effect in a perpendicularly oriented layer can be used
as a light valve.

The transient response behavior of the dielectric alignment
process is of course of interest in the application of this effect
to electro-optical display devices. The corresponding transient
response behavior for the magnetic analog of dielectric alignment
has been treated by Brochard et al. [121] for both the parallel
and perpendicularly oriented nematic films. Considering the
assumptions made, their equations should be applicable to the
electric field case. The solution of this problem involves a

simplification of the Leslie-Ericksen continuum equations (Sec. IIA) using the small angle approximation, among other assumptions, much as was done for the case of Williams domains to get equation (2). If a stepwise voltage change is imposed across the sample, the initial exponential growth or decay of the deformation to its final state is governed by a time constant τ. For the perpendicularly oriented nematic layer

$$\frac{1}{\tau} = \left(\frac{\pi}{x_0}\right)^2 \frac{k_{33}}{c(\alpha_3 - \alpha_2)} \left[\left(\frac{U}{U_{0\varepsilon}}\right)^2 - 1\right] \tag{14}$$

where $U_{0\varepsilon}$ is the threshold voltage given by (13), x_0 is the sample thickness, k_{33} is the bend elastic constant (Sec. IA3), and α_2 and α_3 are two of the five independent Leslie viscosities [70,85]. c is a factor that corrects for the transient backflow effects in the nematic layer. Brochard, Pieranski and Guyon's [121] experiments are in good agreement with the magnetic analog of the $[(U/U_{0\varepsilon})^2 - 1]$ term of Eq. (14). A deformation will be exponentially damped out if $U < U_{0\varepsilon}$ and will grow larger if $U > U_{0\varepsilon}$.

There is very little published data on the response times of the dielectric alignment effect. Schiekel and Fahrenschon [51] and Soref and Rafuse [119] present some transient response data for the perpendicularly oriented film, but there has been no attempt to quantitatively verify the validity of Eq. (14).

C. Flexoelectricity -- Liquid Crystalline Curvature Electricity

Meyer [122] has suggested that certain liquid crystals may show an effect analogous to piezoelectricity in solid crystals where a spatial reorientation of the optic axis in a deformed nematic corresponds to the spatial displacement of the atoms in a solid crystal. Since pressure does not play a role, de Gennes has given Meyer's effect the new name of "flexoelectricity."

Molecules making up the nematic liquid crystalline phase must possess a permanent dipole moment and have an asymmetrical shape to show this effect. Consider, for example, wedgelike molecules in a wedge-shaped container where the boundary surfaces induce a splay deformation as is illustrated in Fig. 3. On the average, more wedge-shaped molecules will orient with their thick ends towards the wide part of cell than towards the narrow part because of their shape. This net orientation results in a nonvanishing macroscopic polarization if the molecules have a component of permanent dipole moment in the direction of their long axes. A similar argument also holds for crescent-shaped molecules in a bend deformation if there is a component of permanent dipole moment perpendicular to the long molecular axis. The reverse effect must also occur; an electric field should induce a splay or a bend deformation in a sample composed of these types of molecules. This is a much weaker effect than the dielectric alignment of a nematic described in Secs. IIIA and IIIB and should only be observable under conditions where pure dielectric alignment is suppressed. Furthermore, it should be only seen in dc fields.

Helfrich [123] has estimated the magnitude of the flexoelectric coefficients for nematic liquid crystals from a consideration of typical molecular sizes, shapes and dipole moments and typical nematic liquid crystal elastic constants.

Helfrich [124a] points out that flexoelectricity should be easily measured by observing the field-induced birefringence in a perpendicularly oriented layer that has electrodes positioned so that the electric field is parallel to the two confining plates. For this experiment it is important that the anchorage of the molecules at the boundaries be not too rigid. This geometry is different from the perpendicular sandwich cell arrangement described in Sec. IIIB. Here, the nematic liquid crystal is confined between two parallel, nonconducting glass plates. Two parallel aluminum strips 15-220 μm thick separated by about 2 mm serve as spacers for the glass plates and are the electrodes as well. The

optical transmission of the sample cell is measured normal to the
glass plates.

Helfrich interprets the results of an experiment of Haas,
Adams and Flannery [49] in this geometry as being caused by
flexoelectricity. Recently, more complete experiments by Schmidt,
Schadt and Helfrich [48] in the same geometry give a value for the
bend flexoelectric coefficient that is in good agreement with the
theoretical estimate. A dc field is applied to the electrodes of
a cell filled with MBBA and the observed increase in the birefringence indicates that the local optic axis tips in a direction
parallel to the field. The birefringence is proportional to E^2,
and there is no threshold field. Their observations eliminate the
possibility of fluid flow, and the deformation cannot be a result
of dielectric alignment because MBBA has a negative $\Delta\varepsilon$.

Under certain circumstances flexoelectricity may enable nematic liquid crystals to show effects corresponding to ferroelectric
domains in solids. A spontaneously deformed structure can become
stable if the flexoelectric energy can become larger than the
ordinary elastic energy. Gruler et al. [124b] show that this condition may be fulfilled near the nematic $\rightarrow$ smectic A phase transition and that is is possible for an arrangement of domains to form
if a negative interfacial tension exists between the two mesophases. Domains have been observed near the nematic $\rightarrow$ smectic A
transition [124b] and these structures agree quite well with the
flexoelectric model.

D. Schadt-Helfrich Twist Cell

If one electrode of a parallel oriented sandwich cell (Sec.
IIIA) is rotated by an angle θ with respect to the other
(typically $0 \leq \theta \leq \pi$), a third important sample geometry for
nematics is generated. With no field present the optic axis of
the nematic is still parallel to the plates but now it turns

uniformly from one electrode to the other. With an applied
electric field a deformation takes place due to dielectric alignment that is more complex than discussed in Sec. IIIA, because
now both the polar angle and the azimuthal angle of the local optic axis orientation come into play. Leslie [125] has computed
the analogous deformation for the case of magnetic fields applied
normal to the electrodes. An exact treatment of the case for
electric fields has not yet been given but Leslie's results should
be a good approximation when $\Delta\varepsilon$ is small. The analogous expression for the threshold voltage $V_{0\varepsilon}$ is valid even for large $\Delta\varepsilon$.

$$V_{0\varepsilon} = \left\{ \frac{\pi^2 k_{11} + (k_{33} - 2k_{22})\theta^2}{\varepsilon_0 \Delta\varepsilon} \right\}^{1/2} \tag{15}$$

If the polarization plane of narmally incident light is either
parallel or perpendicular to the optic axis direction at the
transparent electrode, the polarization plane will rotate exactly
in step with the twisted structure, as long as the wavelength of
the light is small compared with the cell thickness [55]. When a
voltage several times the threshold value is applied, however, the
twisted structure is transformed into a nearly perpendicularly
oriented nematic film which can no longer rotate the light. It is
therefore possible to switch the polarization plane of a light
beam from 0 to 90° when $\theta = \pi/2$ by turning the cell "on" or
"off." If the twisted sandwich cell is placed between crossed or
parallel polarizers, the arrangement will then act as a light
valve [126]. This configuration can be more efficient that light
valves based upon the parallel or perpendicularly oriented nematic
layers. This device also has several advantages in comparison
with the dynamic scattering mode devices discussed in Sec. IIE.
The Schadt-Helfrich twist cell does not consume electrical power
in its field on state because there is no ionic flow in the nematic material. Electrochemical reactions can therefore be avoided
and the lifetime of the device can be increased. Furthermore,

this dielectric alignment effect can operate at a much lower
voltage than the dynamic scattering mode. Some new room temperature nematic materials have recently been developed that have
electro-optical thresholds in the neighborhood of 1 V [127a].
Multicolor displays are also possible [127b].

The dynamic behavior of the twisted nematic deformation
should be similar to the behavior of a parallel or perpendicularly
oriented nematic layer. In particular, a $[(V/V_{0\varepsilon})^2 - 1]^{-1}$
dependence of the response time (Eq. 14) is expected to apply.
Response time measurements on a room temperature nematic mixture
have been made by Jakeman and Raynes [128a] and Jones and Lu
[128b].

E. Cholesteric to Nematic Phase Change

Another variety of electro-optical effects can occur in
cholesteric liquid crystals, because their built-in screw structure adds another degree of freedom. When the repeating distance
of the twisted structure is comparable to the wavelength of visible light, the interference of light in an oriented film produces
vivid selective reflection colors [53-58]. It is possible to alter the pitch of the structure with an electric field, and so in
principle the selective reflection band can be "tuned" across the
visible spectrum.

De Gennes [129] and Meyer [130] first hypothesized that a
magnetic field applied perpendicular to the helical axis of a
cholesteric liquid crystal would increase the pitch by unwinding
the structure. De Gennes gives an exact relationship between the
distorted pitch length and the magnetic field. At a certain
critical field the twisted structure is completely unwound, giving
a nematic structure. A dielectric alignment can occur in an
electric field which is exactly analogous to the magnetic case.
The electric field is applied normal to the helical axis of a
cholesteric material having $\Delta\varepsilon > 0$. This configuration can be

realized in a transparent electrode sandwich cell when the
cholesteric liquid crystal is in its so-called focal conic or
"fingerprint" texture. In this texture the helical axis of the
structure lies parallel to the cell boundaries and perpendicular
to the electric field. In transmitted light the observer sees an
irregular pattern of domains that strongly scatters the light. In
this geometry the critical electric field for complete unwinding
of the helical structure is

$$E_c = \frac{\pi^2}{P_0} \left(\frac{k_{22}}{\varepsilon_0 \Delta \varepsilon} \right)^{1/2} \tag{16}$$

P_0 is the intrinsic pitch and k_{22} is the twist elastic constant. $E_c \cong 10$ V/μm for typical values of $k_{22} = 10^{-6}$ dyn,
$P_0 = 1$ μm and $\Delta \varepsilon = 1$. It should be emphasized that this if a
field effect in contrast to the voltage effects that were
encountered in Secs. IIIA, B, D, because the characteristic
length is now the built-in pitch length of the cholesteric structure and not the thickness of the sample cell. The relationship
between the deformed pitch P and the electric field E can be
given in reduced form in terms of a parameter k ($0 \le k \le 1$).

$$\frac{P}{P_0} = \frac{4E(k)K(k)}{\pi^2}$$
$$\frac{E}{E_c} = \frac{k}{E(k)} \tag{17}$$

$K(k)$ and $E(k)$ are the complete elliptic integrals of the first
and second kinds, respectively. Figure 14 shows a theoretical
curve of P/P_0 vs. E/E_c calculated from Eq. (17). Baessler et
al. [131] and Kahn [132] have experimentally measured the
cholesteric pitch as a function of electric field and find good
agreement with this curve. In Kahn's experiment the liquid crystal was in its focal-conic texture and the pitch and angle of
incidence were of the right value to selectively reflect light
over the entire visible spectrum depending on the applied voltage.

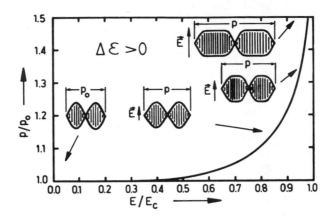

FIG. 14.   Electric field dependence of the cholesteric pitch
shown in reduced form.

De Gennes's calculations show that the optic axis in the distorted
helical structure no longer turns uniformly, as is illustrated in
Fig. 14.   Higher order selective reflections of normally incident
light which are forbidden for a uniformly twisted structure [54]
should be allowed in such a distorted structure.   Chou et al.
[133] have computed the structure and polarization of the first
and second order reflection bands for such a partially unwound
cholesteric liquid crystal, but there has as yet been no experimental verification of their results.   The only experimental check
of the form of such a partially unwound structure comes from the
magnetic susceptibility measurements of Regaya et al. [134a] and
the ESR measurements of Luckhurst and Smith [134b].

The transformation from the cholesteric structure to a
nematic structure for applied fields larger than E_c is a useful
electro-optic phenomenon.   Wysocki et al. [135a] were first to
demonstrate this phase change effect in electric fields.   A display device taking advantage of this effect is illustrated in
Fig. 15a.   With no voltage applied to the cell the random distribution of focal conic regions very efficiently scatters the incident light and the sample appears cloudy.   When a field is

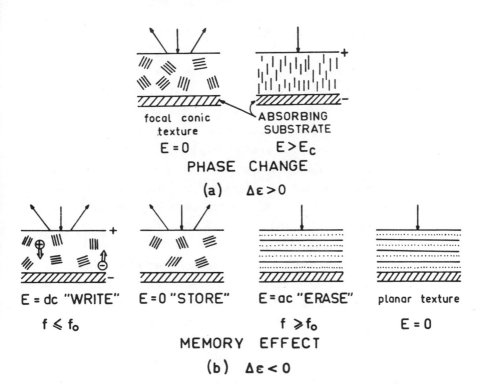

FIG. 15.   (a) A display device based upon the electric field induced phase change from a cholesteric to a nematic liquid crystal.   The upper and lower plates are optically transparent and absorbing electrodes, respectively.   (b) Control sequence for a display device with a memory.

applied greater than E_c, however, the sample is transformed to the transparent nematic state and the light is absorbed by the black substrate. The very sharp decrease in the light scattering of the cholesteric phase at the critical field makes this effect ideally suited for use in matrix displays where the excitation of unselected elements, or crosstalk, has been a problem in conventional nematic displays [135b].   Because it is a true field effect, however, rather substantial voltages in the neighborhod of 100 V must be applied to the electrodes of a practical device that employs this cholesteric phase change effect.

F.   Deformations from the Cholesteric Planar Texture

Electro-optical effects can also occur in the planar
cholesteric texture where the helical axis is uniformly perpendicular to the electrodes.   In contrast to the focal-conic cholesteric texture, the planar texture does not scatter the light.
This texture can usually be produced by introducing the cholesteric liquid crystal between two parallel, rubbed glass plates and
giving them a slight mechanical shear.   If the plates are transparent electrodes, it becomes possible to look along the helical
axis in the direction of the applied field.   For $\Delta\varepsilon < 0$ this
planar texture, illustrated in Fig. 15b, will be stable.

Heilmeier and Goldmacher [136] describe an electro-optic
device having a memory that is based on the transformation between
the focal conic and the planar texture.   This scheme is illustrated in Fig. 15b.   First, a dc or low frequency ac voltage is
applied to a transparent upper electrode and an absorbing lower
electrode that are in contact with the planar cholesteric sample.
Within about 30 msec the sample changes from its transparent
planar texture to a cloudy white focal conic texture that strongly
scatters light.   The uniformly twisted structure is severely disrupted by the turbulent flow induced by the injection and formation of space charges.   This disrupted state can last for days
after the "writing" voltage is removed.   The storage time of this
optical memory effect has been studied by Hulin [137a].   His results indicate that the lifetime of the disrupted texture depends
exponentially upon the ratio of the sample thickness to the
cholesteric pitch.   He was able to change the storage time from
1 minute to several days by varying this ratio from 2 to 15.   The
transparent planar state can again be restored at any time by
applying an ac "erasing" voltage having a frequency higher than
the space charge limited dielectric relaxation frequency f_0.
This is the same erasing scheme use by Melchior et al. [65] in
their thermally addressed display device.   The time for complete

erasure requires from 1/2 to 2 seconds, depending on the applied
field.   This is a pure dielectric alignment process.

Other electro-optical effects can take place in the planar
cholesteric texture when the electric field is applied parallel
to the helical axis and the dielectric anisotropy is positive.
Gerritsma and van Zanten [137b], for example, have observed an
electric field induced shift of the cholesteric selective reflection band to shorter wavelengths.   This situation is much more
difficult to treat theoretically than was the case discussed in
Sec. IIIE where the field was perpendicular to the helical axis.
Leslie [125] has considered the magneto-optical analog where he
assumes that a one dimensional deformation occurs from the planar
cholesteric texture with the local optic axis reorienting only in
a direction parallel to the applied field.   This distortion is
sometimes referred to as the "conical deformation" [129] and
approximate treatments of it have been given in earlier papers
[129,130].   This deformation is similar to the nematic distortions
encountered in Secs. IIIA, B, and D because they are all planar;
i.e., the local optic axis is uniformly oriented on any plane that
is parallel to the boundary surfaces.   Leslie's equations are
valid only if the following condition is fulfilled:

$$\text{Either} \left(\frac{k_{33}}{k_{22}}\right) \leq 1 \quad \text{or} \quad \left(\frac{x_0}{P_0}\right)^2 < \frac{1}{12[(k_{33}/k_{22}) - 1]}$$

If the first condition is satisfied, Leslie shows that the
cholesteric pitch should decrease with increasing field.   This
cannot explain Gerritsma and van Zanten's "blue shift", however,
because $k_{33}/k_{22} > 1$ for all nematic and cholesteric liquid crystals that have had their elastic constants measured so far.
MBBA, for example, has $k_{33}/k_{22} \cong 2.3$ [137c].   Leslie's second
condition is generally fulfilled only if the sample cell contains
less than one complete turn of the cholesteric helix.   No selective reflection of light would be possible under these circumstances, and besides, Leslie's equations predict an increase of

the pitch for this case. The "blue shift" therefore cannot be
explained by a field induced pitch contraction. Gerritsma and
van Zanten [138] have recently proposed a completely new explanation for this effect.

What actually happens when an electric field is applied to a
sample containing many turns of the cholesteric helix and
$k_{33}/k_{22} > 1$ is quite surprising. A three-dimensional deformation
is observed to occur where the planar texture breaks up into a
grid pattern of focals [66-68,139] illustrated in Fig. 16.
Helfrich [86,140] and Hurault [69] have considered the possibility
of nonplanar, two-dimensional distortions from the planar cholesteric texture. In addition to the reorientation of the local optic axis in the direction of the field, they also permit a
periodic reorientation in a direction perpendicular to the applied
field. Their treatment is only valid when there are many turns of
the cholesteric helix in the layer. Hurault's revisions of
Helfrich's expressions lead to a periodic distortion having a
wavelength λ given by

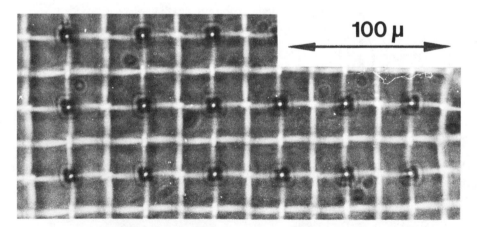

FIG. 16. The grid pattern resulting from the application of
29.1 V at 2 kHz to a cholesteric layer 114 µm thick. The pitch of
the cholesteric helix is 10.6 µm. (Data from Ref. 67.)

$$\lambda^2 = \frac{3}{2} (k_{33}/k_{22})^{1/2} P_0 x_0 \tag{18}$$

In the pure dielectric alignment regime where the frequency of
the applied voltage is much higher than f_0, the threshold voltage $V_{0\varepsilon}$ for the deformation is

$$V_{0\varepsilon}^2 = 2\pi^2 \frac{\varepsilon_1 + \varepsilon_2}{\varepsilon_1 - \varepsilon_2} \frac{1}{\varepsilon_0 \varepsilon_2} \left(\frac{3}{2} k_{22} k_{33} \right)^{1/2} \frac{x_0}{P_0} \tag{19}$$

Helfrich [101] and Hurault [69] also give expressions for the
threshold voltage in the conduction regime. The situation is more
complicated in this regime because space charge and fluid flow
also play a role. The grid pattern in this regime corresponds to
Williams domains in nematic liquid crystals. Predictions for
this regime have been partially verified by experiment [66-68].

Experimental observations made on the three dimensional grid
pattern in the dielectric regime are in very good agreement with
Eqs. (18) and (19) which were derived for two dimensional distortions. Not only has the x_0 and P_0 dependence of these equations been verified [67,68,139], but the elastic constants
obtained from these expressions are in agreement with known values
as well. This agreement indicates that the three-dimensional
pattern can be considered as a superposition of two orthogonal,
two-dimensional patterns when the distortions are weak [140].

REFERENCES

1. W. Kast, Angew. Chem., 67, 592 (1955).

2. G. H. Brown and W. G. Shaw, Chem. Rev., 57, 1049 (1957).

3. I. G. Chistyakov, Soviet Phys. Usp., 9, 551 (1967).

4. A. Saupe, Angew. Chem. Int. Edit., 7, 97 (1968).

5. H. Sackmann and D. Demus, Fortschr. Chem. Forsch., 12, part
    2, 349 (1969).

6. H. Baessler, Festkörperprobleme XI - Advances in Solid State
    Physics (O. Madelung, ed.), 99, Pergamon and Kieweg, 1971.

7.   G. W. Gray, Molecular Structure and the Properties of Liquid
     Cyrstals, Academic, London, 1962; G. W. Gray and P. A.
     Winsor (eds.), Liquid Crystalline and Plastic Crystalline
     Mesophases, Van Nostrand-Reinhold, 1974; P. G. de Gennes, The
     Physics of Liquid Crystals, Oxford University Press, London,
     1974.

8.   R. W. Gurtler and C. Maze, IEEE Spectrum, Nov. 25 (1972);
     A. Sussman, IEEE Trans. on Parts, Hybrids and Packaging
     PHP-8, 24 (1972).

9.   M. Schadt and W. Helfrich, Mol. Cryst. Liq. Cryst., 17, 355
     (1972).

10. W. J. Toth and A. V. Tobolsky, Polymer Letters, 8, 531
     (1970).

11. J. B. Stamatoff, Mol. Cryst. Liq. Cryst., 16, 137 (1972).

12. V. Luzzati, H. Mustacchi, A. E. Skoulios, and F. Husson, Acta
     Cryst., 3, 660 (1960).

13. P. A. Spegt and A. E. Skoulios, Acta Cryst., 17, 198 (1964).

14. A. S. C. Lawrence, Mol. Cryst. Liq. Cryst., 7, 1 (1969).

15. A. Elliott and E. J. Ambrose, Disc. Faraday Soc., 9, 246
     (1950).

16. C. Robinson, Mol. Cryst., 1, 467 (1966).

17. G. T. Stewart, Mol. Cryst., 1, 563 (1966); Adv. Chem. Ser.,
     63, 141 (1967); Mol. Cryst. Liq. Cryst., 7, 75 (1969).

18. G. Friedel, Ann. Phys., 18, 273 (1922).

19. D. Demus, S. Diele, M. Klapperstück, V. Link, and H.
     Zaschke, Mol. Cryst. Liq. Cryst., 15, 161 (1971).

20. The various known material constants for PAA were extracted
     from the literature and collected in Refs. 72, 73, and 84a.

21. H. Kelker and B. Scheurle, J. Phys., 30, C4-104 (1969);
     Angew. Chem., 81, 903 (1969).

22. The material constants for MBBA have been collected by
     P. A. Penz and G. W. Ford in Ref. 72.

23a. J. A. Castellano, RCA Rev., 33, 296 (1972); Ferroelectrics,
     3, 29 (1971).

23b. G. W. Gray, K. J. Harrison, and J. A. Nash, Electron.
     Letters, 9, 130 (1973).

24. J. Nehring and A. Saupe, J. Chem. Soc., Faraday Trans. II,
     68, 1 (1972).

25. A. Saupe and W. Maier, Z. Naturforsch., 16a, 816 (1961).

26. W. Maier and A. Saupe, Z. Naturforsch., 14a, 882 (1959);
     15a, 287 (1960).

27. A. Saupe, Mol. Cryst. Liq. Cryst., 16, 87 (1972).

28. D. A. Balzerini, Phys. Rev. Letters, 25, 914 (1970).

29. I. Haller, H. A. Huggins, and M. J. Freiser, Mol. Crys. Liq. Cryst., 16, 53 (1972).

30. W. Maier and G. Meier, Z. Naturforsch., 16a, 262 (1961); 16a, 470 (1961).

31. M. Schadt, J. Chem. Phys., 56, 1494 (1972).

32. W. H. De Jeu, C. J. Gerritsma, P. van Zanten, and W. J. A. Goossens, Phys. Letters, 39A, 355 (1972).

33. A. Martin, G. Meier, and A. Saupe, Symp. Faraday Soc., 5, 119 (1971).

34. F. Rondelez, Solid State Commun., 11, 1675 (1972).

35. T. Svedberg, Ann. Physik, 44, 1121 (1914); 49, 437 (1916).

36. F. Rondelez, D. Diguet and G. Durand, Mol. Crys. Liq. Cryst., 15, 183 (1971).

37. R. Twitchell and E. F. Carr, J. Chem. Phys., 46, 2765 (1967).

38. H. Zocher, Z. Physik, 28, 790 (1927).

39. C. W. Oseen, Ark. Mat. Astron. Fys., A19, 1 (1925); Fortschr. Chem. Phys. Phys. Chem., 20, 1 (1929); Trans. Faraday Soc., 29, 883 (1933).

40. F. C. Frank, Disc. Faraday Soc., 25, 19 (1958).

41. J. Nehring and A. Saupe, J. Chem. Phys., 54, 337 (1971).

42. A. Saupe, Z. Naturforsch., 15a, 815 (1960).

43. A review on elastic constant measurements is given by H. Gruler, T. J. Scheffer, and G. Meier, Z. Naturforsch., 27a, 966 (1972).

44. I. Haller, J. Chem. Phys., 57, 1400 (1972).

45. H. Gruler, Z. Naturforsch., 28a, 474 (1973); H. Gruler and G. Meier, Mol. Cryst. Liq. Cryst., 23, 261 (1973).

46. P. Chatelain, Bull. Soc. Franc. Mineral., 66, 105 (1943).

47a. D. W. Berreman, Phys. Rev. Letters, 28, 1683 (1972).

47b. J. L. Janning, Appl. Phys. Letters, 21, 173 (1972).

48. D. Schmidt, M. Schadt and W. Helfrich, Z. Naturforsch., 27a, 277 (1972).

49. W. Haas, J. Adams, and J. B. Flannery, Phys. Rev. Letters, 25, 1236 (1970).

50. P. E. Cladis, J. Rault, and J. P. Burger, Mol. Cryst. Liq. Cryst., 13, 1 (1971).

51. M. F. Schiekel and K. Fahrenschon, Appl. Phys. Letters, $\underline{19}$, 391 (1971).

52. J. E. Proust and L. Ter-Minassian-Saraga, Solid State Commun., $\underline{11}$, 1227 (1972).

     F. J. Kahn, Appl. Phys. Letters, $\underline{22}$, 386 (1973).

53. D. W. Berreman and T. J. Scheffer, Phys. Rev., $\underline{A5}$, 1397 (1972).

54. C. W. Oseen, Trans. Faraday Soc., $\underline{29}$, 833 (1933).

55. Hl. De Vries, Acta Cryst., $\underline{4}$, 219 (1951).

56. D. W. Berreman and T. J. Scheffer, Phys. Rev. Letters, $\underline{25}$, 577 (1970); Mol. Cryst. Liq. Cryst., $\underline{11}$, 395 (1970); D. W. Berreman, J. Opt. Soc. Am., $\underline{62}$, 502 (1972).

57. R. Dreher, G. Meier, and A. Saupe, Mol. Cryst. Liq. Cryst., $\underline{13}$, 17 (1971).

58. R. Dreher and G. Meier, Phys. Rev., $\underline{A8}$, 1616 (1973).

59. J. Seelig, J. Am. Chem. Soc., $\underline{92}$, 3881 (1970).

60. V. Fréedericksz and A. Repiewa, Acta Physicochim. URSS, $\underline{4}$, 91 (1936).

61. L. K. Vistin' and A. P. Kapustin, Sov. Phys. Crystallog., $\underline{13}$, 284 (1968).

62. J. A. Geurst and W. J. A. Goossens, Phys. Letters, $\underline{41}$, 369 (1972).

63. E. F. Carr, Mol. Cryst. Liq. Cryst., $\underline{13}$, 27 (1971).

64a. C. Tani, Appl. Phys. Letters, $\underline{19}$, 241 (1971).

64b. F. J. Kahn, Appl. Phys. Letters, $\underline{22}$, 111 (1973).

65. H. Melchior, F. J. Kahn, D. Maydan, and D. B. Fraser, Appl. Phys. Letters, $\underline{21}$, 392 (1972).

66. F. Rondelez and H. Arnould, C. R. Acad. Sci., $\underline{B273}$, 549 (1971).

67. T. J. Scheffer, Phys. Rev. Letters, $\underline{28}$, 593 (1972).

68. F. Rondelez, H. Arnould, and C. J. Gerritsma, Phys. Rev. Letters, $\underline{28}$, 735 (1972).

69. J. P. Hurault, J. Chem. Phys., $\underline{59}$, 2068 (1973).

70. F. M. Leslie, Quart. J. Mech. Appl. Math., $\underline{19}$, 357 (1966); Arch. Rational Mech. Anal., $\underline{28}$, 265 (1968).

71. J. L. Ericksen, Arch. Rational Mech. Anal., $\underline{9}$, 371 (1962); Mol. Cryst. Liq. Cryst., $\underline{7}$, 153 (1969). The latter paper is a review of the Leslie-Ericksen theory and contains many earlier references.

72. P. A. Penz and G. W. Ford, Phys. Rev., A6, 414, 1676 (1972).

73. H. C. Tseng, D. L. Silver, and B. A. Finlayson, Phys.
     Fluids, 15, 1213 (1972).

74. E. W. Aslaksen, Phys. Kondens. Mater., 14, 80 (1971).

75. E. Dubois-Violette, P. G. de Gennes, and O. Parodi, J.
     Phys., 32, 305 (1971).

76. R. Williams, J. Chem. Phys., 39, 384 (1963).

77. P. A. Penz, Phys. Rev. Letters, 24, 1405 (1970); Mol. Cryst.
     Liq. Cryst., 15, 141 (1971). The latter paper contains a
     brief historical review.

78. P. Kassubek, Diplomarbeit Universität Freiburg, W. Germany.

79. W. Helfrich, J. Chem. Phys., 51, 2755 (1969).

80. J. Nehring, Unpublished results.

81. C. Deutsch and P. N. Keating, J. Appl. Phys., 40, 4049
     (1969); G. Assouline, A. Dmitrieff, M. Hareng, and E. Leiba,
     C. R. Acad. Sci., B271, 857 (1970); E. W. Aslaksen and
     B. Ineichen, J. Appl. Phys., 42, 882 (1971). In a different
     geometry: W. W. Holloway, Jr., and M. J. Rafuse, J. Appl.
     Phys., 42, 5395 (1971).

82. T. O. Carroll, J. Appl. Phys., 43, 767 (1972).

83. E. F. Carr, Mol. Cryst. Liq. Cryst., 7, 253 (1969).

84a. W. Helfrich, J. Chem. Phys., 51, 4092 (1969).

84b. H. Gruler, Mol. Cryst. Liq. Cryst., 27, 31 (1974).

85. Relationships between these viscosity parameters and those
     used by other authors are given, for example, by M.
     Papoular, Phys. Letters, 30A, 5 (1969). The quantity
     $1/2 (\alpha_4 - \alpha_2 + \alpha_5)$ is the viscosity that would be measured
     when the optic axis is parallel to the velocity gradient.

86. This proportionality constant represents an average value
     taken from Ref. 76 and 77.

87. Penz's [77] empirical relation $V = (5.5 \pm 0.2)(\lambda/2x_0)$ volts
     was used to compute this value. It should be noted that this
     is a low frequency ac rms value, while Helfrich's theory is
     strictly valid only for dc.

88. L. K. Vistin', Sov. Phys. Crystallog., 15, 514 (1970).

89. W. Greubel and U. Wolff, Appl. Phys. Letters, 19, 213 (1971).

90. Orsay Liquid Crystal Group, Phys. Rev. Letters, 25, 1642
     (1970).

91. Orsay Liquid Crystal Group, Mol. Cryst. Liq. Cryst., 12, 251
     (1971).

92. G. H. Heilmeier and W. Helfrich, Appl. Phys. Letters, 16, 155
     (1970).

93. Y. Galerne, G. Durand, and M. Veyssié, Phys. Rev., A6, 484
     (1972).

94. Y. Galerne, G. Durand, M. Veyssié, and V. Pontikis, Phys.
     Letters, 38A, 449 (1972).

95. E. Dubois-Violette, J. Phys., 33, 95 (1972).

96. Orsay Liquid Crystal Group, Phys. Letters, 39A, 181 (1972).

97. W. H. De Jeu, Phys. Letters, 37A, 365 (1971).

98. A comparative study of dc and ac excitation is given in Ref.
     91.

99. H. Koelmans and A. M. van Boxtel, Phys. Letters, 32A, 32
     (1970); Mol. Cryst. Liq. Cryst., 12, 185 (1971).

100. N. Felici, Rev. Gen. Elec., 78, 717 (1969).

101. W. Helfrich, J. Chem. Phys., 55, 839 (1971).

102. P. G. de Gennes, Comments Solid State Phys., 3, 35, 148
     (1970).

103. G. Durand, M. Veyssié, F. Rondelez, and L. Leger, C. R.
     Acad. Sci., B270, 97 (1970).

104. Y. Björnstahl, Ann. Physik, 56, 161 (1918).

105. G. H. Heilmeier, L. A. Zanoni, and L. A. Barton, Proc. IEEE,
     56, 1162 (1968); IEEE Trans., ED-17, 22 (1970).

106. L. T. Creagh, A. R. Kmetz, and R. A. Reynolds, IEEE Trans.
     ED-18, 672 (1971). This paper includes many experimental
     results and gives a good overall view of the dynamic
     scattering mode.

107. T. O. Carroll, J. Appl. Phys., 43, 1342 (1972).

108. P. J. Wild and J. Nehring, Appl. Phys. Letters, 19, 335
     (1971).

109. C. R. Stein and R. A. Kashnow, Appl. Phys. Letters, 19, 343
     (1971). J. G. Grabmaier, W. F. Gruebel, and H. H. Krüger,
     Mol. Cryst. Liq. Cryst., 15, 95 (1971).

110. J. D. Margerum, J. Nimoy, and S.-Y. Wong, Appl. Phys. Letters,
     Letters, 17, 51 (1970; T. D. Beard, W. P. Bleha, Jr., and
     S.-Y. Wong, Appl. Phys. Letters, 22, 90 (1973); D. L. White
     and W. Feldman, Electron. Letters, 6, 837 (1970).
     G. Assouline, M. Hareng, and E. Leiba, Proc. IEEE, 59, 1355
     (1971).

111a. H. Kiemle and U. Wolff, Optics Commun., 3, 26 (1971);
     J. D. Margerum, T. D. Beard, W. P. Bleha, Jr. and S.-Y. Wong,
     Appl. Phys. Letters, 19, 216 (1971).

111b. H. Gruler and L. Cheung, J. Appl. Phys., 46, 5097 (1975).

112.   Pioneering work done by H. Zocher and the Russian group is
       summarized in the following references: H. Zocher, Trans.
       Faraday Soc., 29, 945 (1933); V. Fréederichsz and
       Z. Kristallog., 79, 225 (1931); V. Fréederichsz and
       V. Zwetkoff, Sov. Phys., 6, 490 (1934); V. Zwetkoff, Acta
       Physicochim. URSS, 6, 865 (1937).

113.   C. M. Dafermos, SIAM J. Appl. Math., 16, 1305 (1968).

114.   A. Rapini and M. Papoular, J. Phys., 30, C4-54 (1969).

115.   H. Gruler and G. Meier, Mol. Cryst. Liq. Cryst., 16, 229
       (1972).

116a. The occurrence of a threshold value, below which absolutely
       no deformation can occur is nearly analogous to the engineer-
       ing problem of the loaded column, which states that if a
       compressive force acts exactly along the axis of a uniform
       rod, a critical force exists above which the rod will curve
       or buckle sideways. See, for example, R. A. Beth, in
       Handbook of Physics (E. U. Condon and H. Odishaw, eds.),
       McGraw-Hill, New York, 1958, pp. 3-72 to 3-75.

116b. L. Leger, Solid State Comm., 11, 1499 (1972).

116c. H. J. Deuling, Mol. Cryst. Liq. Cryst., 19, 123 (1972).

117.   G. H. Heilmeier and L. A. Zanoni, Appl. Phys. Letters, 13, 91
       (1968).

118.   F. J. Kahn, Appl. Phys. Letters, 20, 199 (1972).

119.   R. A. Soref and M. J. Rafuse, J. Appl. Phys., 43, 2029
       (1972).

120.   G. Assouline, M. Hareng, and E. Leiba, Electron. Letters, 7,
       699 (1971); G. Assouline, M. Hareng, E. Leiba, and
       M. Roncillat, Electron. Letters, 8, 45 (1972); M. Hareng,
       E. Leiba, and G. Assouline, Mol. Cryst. Liq. Cryst., 17, 361
       (1972); M. Hareng, G. Assouline, and E. Leiba, Proc. IEEE,
       60, 913 (1972).

121.   F. Brochard, P. Pieranski, and E. Guyon, Phys. Rev. Letters,
       28, 1681 (1972); P. Pieranski, F. Brochard, and E. Guyon,
       J. Phys., 34, 35 (1973).

122.   R. B. Meyer, Phys. Rev. Letters, 22, 918 (1969).

123.   W. Helfrich, Z. Naturforsch., 26a, 833 (1971).

124a. W. Helfrich, Phys. Letters, 35A, 393 (1971).

124b. H. Gruler, L. Cheung, and R. B. Meyer, to be published.

125.   F. M. Leslie, Mol. Cryst. Liq. Cryst., 12, 57 (1970).

126.   M. Schadt and W. Helfrich, Appl. Phys. Letters, 18, 127
       (1971).

127a. A. Boller, H. Scherrer, M. Schadt, and P. J. Wild, Proc. IEEE, 60, 1002 (1972); A. Ashford, J. Constant, J. Kirton, and E. P. Raynes, Electron. Letters, 9, 118 (1973).

127b. T. J. Scheffer, J. Appl. Phys., 44, 4799 (1973).

128a. E. Jakeman and E. P. Raynes, Phys. Letters, 39A, 69 (1972).

128b. D. Jones and S. Lu, 1972 SID Symposium, Digest of Technical Papers, p. 100.

129. P. G. de Gennes, Solid State Comm., 6, 163 (1968).

130. R. B. Meyer, Appl. Phys. Letters, 12, 281 (1968).

131. H. Baessler, T. M. Laronge, and M. M. Labes, J. Chem. Phys., 51, 3213 (1969).

132. F. J. Kahn, Phys. Rev. Letters, 24, 209 (1970).

133. S. C. Chou, L. Cheung, and R. B. Meyer, Solid State Comm., 11, 977 (1972).

134a. B. Regaya, H. Gasparoux, and J. Prost, Rev. Phys. Appl., 7, 83 (1972).

134b. G. Luckhurst and H. Smith, Mol. Cryst. Liq. Cryst., 20, 319 (1973).

135a. J. J. Wysocki, J. Adams, and W. Haas, Phys. Rev. Letters, 20, 1024 (1968).

135b. T. Ohtsuka, M. Tsukamoto, and M. Tsuchiya, Jap. J. Appl. Phys., 12, 371 (1973).

136. G. H. Heilmeier and J. Goldmacher, Appl. Phys. Letters, 13, 132 (1968).

137a. J. P. Hulin, Appl. Phys. Letters, 21, 445 (1972).

137b. C. Gerritsma and P. van Zanten, Mol. Cryst. Liq. Cryst., 15, 257 (1971).

137c. C. Williams and P. E. Cladis, Solid State Comm., 10, 357 (1972).

138. C. J. Gerritsma and P. van Zanten, Phys. Letters, 42A, 329

139. C. J. Gerritsma and P. van Zanten, Phys. Letters, 37A, 47 (1971).

140. W. Helfrich, Appl. Phys. Letters, 17, 531 (1970).

AUTHOR INDEX

Numbers in parentheses are reference numbers and indicate that an author's work is referred to although his name may not be cited in the text. Underlined numbers give the page on which the complete reference is listed.

Part 1, pages 1-528; Part 2, pages 529-818

A

Abbott, A., 493, 503
Abe, Y., 178, 204
Abragam, A., 28(10), 59, 334(27), 365
Abraham, H., 20, 25
Abramowitz, M., 180(39), 205
Abrams, A., 567(12), 598
Adamov, M. N., 494(10), 503
Adams, J., 771(49), 802, 806(175a), 813, 818
Adler, E., 448(42), 455(42), 468
Agranovich, V. M., 398(38), 438, 448(37), 450(37), 468
Akhmanov, S. A., 393(14,26), 436, 437
Akutsu, H., 673(120a), 682
Alberts, B. M., 646(17), 678
Alberty, R. A., 330(21), 365
Alexander, M. H., 475(175), 506
Alfano, R. R., 393(15), 408(69), 435(105), 436, 440, 442
Allais, M. L., 651, 679
Allen, F. S., 91(79a), 118, 251 (33), 253(42), 254(42), 258 (33), 264(33), 265(33), 271, 696, 709
Allen, L. C., 476, 499
Allen, S., 727(62), 728, 740
Alpher, R. A., 508(151), 505
Altick, P. L., 508(159), 505
Ambrose, E. J., (15), 812
Amos, A. T., 484(48), 494, 501, 503
Ananthanarayanan, S., 629, 642
Anderson, J. A., 676(137), 683

Anderson, J. E., 196, 205
Anderson, T. F., 723(57), 740
Andreyeva, L. N., 654(74,78), 663(74), 680
Andriashivili, I. A., 721(36), 739
Angus, J. C., 327(19), 365
Applequist, J., 50, 54(35), 60, 488, 501
Applequist, J. B., 84(65), 118
Arai, T., 484(38), 492(38), 500
Arecci, F. T., 326(11), 365
Armstrong, C. M., 758(34), 760
Arnould, H., 774(66,68), 811(66, 68), 814
Aroney, M. J., 175, 204
Arrighini, G. P., 428(95), 441, 492, 498, 512(94,167), 503, 504, 506
Asai, H., 630, 631, 642
Ashford, A., 804(127a), 818
Aslaksen, E. W., 775(74), 779(81), 789(81), 815
Aspnes, D. E., 386(11), 389
Assouline, G., 779(81), 789(81), 790(110), 799(120), 815, 816, 817
Astbury, W. T., 647, 679
Atanasoff, J. V., 484, 500
Atkins, P. W., 393(21,25), 408(76), 435(110), 437, 440, 443, 447 (28), 455, 467, 468
Auston, D. H., 435(105), 442
Azzi, A., 758, 760

819

SUBJECT INDEX

849